A to Z
Activity Book

Macmillan
McGraw-Hill

Instruction For Copying

Answers are printed in non-reproducible blue. Copy pages on a light setting in order to make multiple copies for classroom use.

Table of Contents

How to Use the A to Z Activity Book

This component provides fun and interesting cross-curricular activities to support science instruction in your Kindergarten classroom. The alphabet is the foundation for these activities, which focus on developing reading, writing, math, and language skills. You will also find art, cooking, music, and movement activities that provide children with opportunities to develop their fine and gross motor skills.

Use this book as a supplement to your science curriculum, and add a new depth to children's learning and understanding of science. You can refer to the *Macmillan/McGraw-Hill Science* Teacher's Edition for suggestions on when to use these activities. The activity content covers a range of areas from life science to earth science to physical science. This component was designed to excite children's natural interest in science and to make teaching science concepts easy for you. Enjoy!

A Special Note about Safety

Some of the activities in this book feature materials which may cause allergic reactions in children, parents, aides, or other school personnel. We encourage you to review the suggested materials for potential allergens before you begin these activities with your class, substituting items as necessary.

 # is for Amazing Ants

Reading — Ant Story

 WHOLE CLASS

Objective: Understand the size of an ant in relation to people.

Science Inquiry Skills: observe, predict, draw a conclusion

Resources: *Two Bad Ants* by Chris Van Allsburg, Graph Head pp. 54–55

- Read the story *Two Bad Ants*. Discuss the illustrations. The children will feel big when they realize that, to an ant, they are very large.

- Copy the Graph Head and make a vertical graph on chart paper. Have children predict how many ants will fit in their hand. Have them write their name and prediction on the prediction side of the graph.

- Then have each child trace their hand on paper and cut it out. Use ant stickers or an ant stamp to put on the hand cutouts. Have children count how many ants fit in their hand. Have them record the number on the conclusion side of the graph.

Writing — An Ant's Body

 WHOLE CLASS

Objective: Learn the parts of an ant's body.

Science Inquiry Skills: classify, make a model

- Discuss with children the three body parts of an ant (head, thorax, abdomen). Help children fold a piece of tan paper in half horizontally and then fold it in thirds the opposite way.

- Help children to cut out three ovals from red paper. Then have them glue an oval on each box as shown.

- Invite children to draw an eye and two antennas on the first oval and six legs coming from the bottom of the middle oval. Then, help children cut up the fold between each box on one side of the tan paper.

- Have children write the body part under each tab: head, thorax, abdomen.

Ant Counting

WHOLE CLASS

Materials
butcher paper, finished products from "An Ant's Body" activity

Objective: Use ant body parts to count by threes.

Science Inquiry Skills: put things in order

- Display children's finished ant body parts on a hill-shaped piece of butcher paper to make a number extension chart for counting by threes.

- Help children to count aloud by threes using the bulletin board display as a guide.

Art

Ant Facts

SMALL GROUP

Materials
white paper, stapler, markers

Objective: Review what children have learned about ants.

Science Inquiry Skills: communicate, draw a conclusion

- Make a 4-page "Amazing Ants" horizontal flipbook, as shown, for each child. Write the text along the bottom of each page.

- Give copies to children and invite them to illustrate and record all the facts they have learned about ants, such as: ants have 6 legs and 3 body parts; live in tunnels underground; can carry big things.

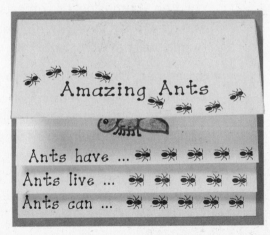

B is for Blowing Bubbles

Science — Bubble Brainstorm

Materials
chart paper, small cups, straws, water, dishwashing liquid

Objective: Learn that bubbles cannot be made from water alone.

Science Inquiry Skills: communicate, observe, compare

- Encourage children to name places where they might see or make bubbles. Record children's responses on chart paper.

- Give children a cup of water and a straw. Have them blow through the straw in the water. Invite children to describe what happens.

- Then add a few drops of dishwashing liquid to each cup and mix with the water. Have children blow through the straw again. Ask children if there was a difference between the cup with only water and the cup with soap and water. Encourage them to explain why.

Math — Blowing Bubbles

Materials
shell, ring, safety pin, pencil, spatula, containers, bubble solution (1 gallon water, $\frac{2}{3}$ cup dishwashing liquid, 1 tbs. glycerin)

Objective: Identify objects that can make bubbles and objects that cannot make bubbles.

Science Inquiry Skills: predict, investigate, conclude

Resources: Graph Head pp. 54–55

- Copy the Graph Head and prepare a vertical picture graph on chart paper. Draw pictures of objects that will and will not make bubbles along the left side of the graph.

- Discuss with children what kinds of objects can be used to make bubbles.

- On the prediction side of the graph, draw a happy face with children's names next to objects they predict will make bubbles and draw a sad face with their names next to objects they predict will not make bubbles.

- Experiment with each object to see which makes bubbles. Record the results on the conclusion side of the graph, again using happy and sad faces.

Can It Make Bubbles?		
	Prediction	Conclusion
	☹	☹
	☺	☺
	☹	☹

Bubble Experiments

Materials

objects with and without holes, containers of bubble solution

Objective: Experiment with objects to find out if they make bubbles.

Science Inquiry Skills: predict, investigate, compare

Resources: "Can It Make Bubbles?" Recording Sheet p. 56

- Have children choose six objects and draw a picture of each on the recording sheet.

- Then ask children to predict if each object will make bubbles or not. Have children record their predictions by drawing a happy or a sad face.

- Invite children to experiment with each object and the bubble solution. Have children record results by drawing a happy or a sad face for each object in the results column of their recording sheet.

Art

Beautiful Bubbles

Materials

cardboard, small bowls, food coloring, dishwashing liquid, water, straws, white paper, markers

Objective: Make bubbles to use to decorate drawings.

Science Inquiry Skills: observe

- Cut out three cardboard patterns of different types of seashells for children to use as tracers.

- Using one bowl for each color, mix food coloring with dishwashing liquid and water. The more food coloring you put in, the brighter the color.

- Help children trace a seashell pattern on white drawing paper using a thick black marker.

- Then have children blow into the colored solution with a straw. Explain that each child should use his/her own straw. When bubbles reach the top of the bowl, have children place their tracings face down on the bubbles and lift the paper up to see the design. Repeat with different colors.

C is for Creating Colors

Reading

Reading About Colors

WHOLE CLASS

Materials
3 clear jars; water; red, blue, and yellow food coloring

Objective: Create new colors from the three primary colors.

Science Inquiry Skills: observe, communicate, predict

Resources: *Mouse Paint* by Ellen Stoll Walsh

- Before reading the story *Mouse Paint*, show children the cover of the book and flip through the pages. Invite children to share their predictions about the story. Encourage children to discuss how they think different colors are made.

- Read the story, using the illustrations to generate discussion about different colors. After reading, explain to children that you are going to make colors.

- Fill 3 clear jars with water. Each time you get ready to create a new color, ask children to predict what it will be. Put red and yellow food coloring in one jar to demonstrate how to make orange. Continue with red/blue and blue/yellow.

Music

Color Song

WHOLE CLASS

Materials
sentence strips; red, yellow, and blue paint; craft sticks; white tag board; red, yellow, and blue cellophane; glue

Objective: Create new colors from the three primary colors.

Science Inquiry Skills: observe, compare, investigate

Resources: Mixing Colors Song p. 57, Crayon Cut-Outs p. 58, Color Mixing Sheet p. 59, Color Paddles p. 60

- Make pocket chart sentence strips with the words to the song. Invite children to place the appropriate colored crayon cut-out in the pocket chart where necessary.

- Then invite children to explore creating new colors using paints and craft sticks on the Color Mixing Sheet.

- For continued support, have children make three color paddles by gluing red, yellow, and blue cellophane to the middle of the paddle shapes cut from tag board. Have children lay two paddles together to make a new color.

Separate Colors

SMALL GROUP

Materials
red and blue candle oil, water, food coloring, 3 small plastic soda bottles, tape

Objective: Observe colors as they mix together and then separate.

Science Inquiry Skills: observe, communicate

- Put red candle oil in a small plastic soda bottle and add water with yellow food coloring. Tape the lid on tightly. Have children shake the bottle to make orange and then watch the colors separate as the liquid settles.

- Repeat using red candle oil with blue water and blue candle oil with yellow water in the remaining two bottles. Invite children to shake, observe, and describe what happens.

Color Quilt

WHOLE CLASS

Materials
purple, orange, and green construction paper; butcher paper; glue; scissors; red, yellow, and blue copy paper

Objective: Create a color quilt.

Science Inquiry Skills: classify, observe, communicate

Resources: Crayons for Color Quilt p. 61

- Cut 7" by 7" squares from purple, green, and orange construction paper. Give each child one square.

- Duplicate the crayons on red, yellow, and blue paper. Ask children to find the two crayon cut-outs that make the color they have and glue them onto the square.

- Create a quilt, alternating colors of squares, by gluing them on a large piece of butcher paper.

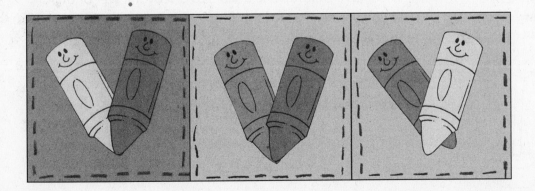

D is for Digging in Dirt

Digging into Dirt

WHOLE CLASS

Objective: Gather and observe soil.

Science Inquiry Skills: observe, investigate

Resources: *Dirt* by Steve Tomecek and Nancy Woodman

Materials

gallon-size plastic bags, small shovels or large spoons, magnifying lenses, strainers, toothpicks

- Read *Dirt* by Steve Tomecek and Nancy Woodman.

- Take children outside. Provide them with small shovels and/or large spoons and plastic bags. Have children dig up some soil and place it in their bags. You can also have each child bring in a small bag of soil from home.

- When you return to the classroom, have children lay out the dirt on paper towels. Invite children to explore the dirt by smelling it, feeling it, looking at it through a magnifying lens, and using a strainer and toothpicks to separate the dirt.

- Have children wash their hands after their explorations.

Math

Weighing Dirt

SMALL GROUP

Objective: Measure and compare the weights of wet and dry soil.

Science Inquiry Skills: measure, predict, compare, communicate

Resources: Graph Head pp. 62–63

wet   dry

Materials

clear plastic cups, dirt, water, scale, paper towels

- Put equal amounts of dry and wet dirt in clear plastic cups. Create a graph on chart paper using the Graph Head.

- Have children predict which cup will weigh more. Record their ideas on the prediction side of the graph.

- Then have children use a scale to discover if wet or dry dirt weighs more. Record the results on the conclusion side of the graph.

- Ask children how the wet soil can turn back to dry soil. Allow the mixture to sit out uncovered overnight or spread it on a towel and place it in a sunny area.

Dirt Dessert

WHOLE CLASS

Objective: Create a fun food item that looks like dirt.

Science Inquiry Skills: measure, make a model

- Have children help to make instant chocolate pudding, following the directions on the package.

- Give each child a bag with cookies. Have children crush the cookies with a rolling pin or their hands and then stir the crumbs into their pudding. Add gummy worms and enjoy the "dirt" dessert.

- Explain to children that the dessert is called "dirt" because it looks like dirt, but it is actually not dirt. Remind children that they cannot eat real dirt.

Pine Cone Forest

SMALL GROUP

Objective: Learn that dirt is necessary for plant growth.

Science Inquiry Skills: observe, compare, communicate

- Have children dip the pine cones in water and then roll one of them in dirt. Sprinkle grass seeds on the pine cones and then place them on a paper towel in a sunny place.

- Provide a spray bottle for children to water the seeds regularly.

- Have children observe and compare the growth of the seeds. Record their observations weekly. Ask: **What do seeds need in order to grow?**

 **is for Extraordinary Eggs**

Reading

Match Book Riddles

SMALL GROUP

Materials
construction paper,
glue, crayons, scissors

Objective: Learn various life forms that hatch from an egg.

Science Inquiry Skills: infer, communicate

Resources: *Chickens Aren't the Only Ones* by Ruth Heller, Animal Sheet p. 64, Cracked Egg Sheet p. 65, Riddle Sheet p. 66

- After reading *Chickens Aren't the Only Ones*, encourage children to brainstorm animals that hatch from eggs. Record children's ideas.

- Prepare a "match book" as shown using the resource pages listed and a $4\frac{1}{2}$" x 12" strip of construction paper. Cut out animals and glue them under the flap of the matching riddle.

- Read the riddles to children and encourage them to guess the animal. Check children's guesses by lifting the flap to reveal the correct animal.

- To use this activity at a center, make match books for each child. Have children find the animal that matches the riddle, color, and glue it under the matching flap.

Writing

The Extraordinary Egg

INDIVIDUAL

Materials
large plastic egg, tote bag, permanent marker, butcher paper

Objective: Involve families in children's writing.

Science Inquiry Skills: predict, communicate

Resources: *An Extraordinary Egg* by Leo Lionni, Family Letter p. 67

- On a large plastic egg, write the following in permanent marker: "What will hatch today?"

- Put the plastic egg and a copy of *An Extraordinary Egg* in a tote bag. You will also need to include a copy of the Family Letter each time a child takes the bag home.

- When the egg is returned, have the child read the clues that they wrote to describe what will "hatch" from the egg. Invite the other children to make guesses.

- Add each item to a large piece of butcher paper cut in an egg shape at your Science Center entitled "That's One Extraordinary Egg!"

Math | The Ever-Floating Egg

SMALL GROUP

Materials
4 clear plastic cups, water, salt, cornstarch, vinegar, red food coloring, 4 eggs, spoon

Objective: Predict and observe which liquids eggs float in.

Science Inquiry Skills: predict, observe, investigate

Resources: Egg Recording Sheet p. 68, Graph Head pp. 62–63

- Explain to children that they will explore what makes uncooked eggs float.

- Give each child a recording sheet. Have children record their predictions. Place the Graph Head on a piece of chart paper and record children's predictions in the left column of the graph.

- Discuss the graph with children, counting the predictions in each column. Ask: **Which has more? Which has less? Which has the same?**

- Conduct the experiment and discuss the results.

- Have children complete the recording sheet and record their results on the conclusion side of the graph. Discuss the change in the graph, who made the correct prediction, and why.

Art | Egg-cellent Egg Art

SMALL GROUP

Materials
8" x 8" squares of construction paper, squeeze bottle, black tempera paint, glue, watercolors, crayons, paper scraps, cardboard patterns of 1 large egg and 1 small egg, markers

Objective: Assess what children have learned about what hatches from an egg.

Science Inquiry Skills: communicate, classify

- In the squeeze bottle, mix 1 part black tempera paint and 1 part white glue. Have children trace the large egg pattern onto white paper. Invite children to make lines on their egg with the squeeze bottle. Let eggs dry overnight.

- When glue and paint dry, have children paint the inside of the egg with watercolors. When the watercolor dries, help children cut out the egg shape and glue it on a square.

- Have children trace and cut out the small egg shape on white paper and glue it on a square. Encourage children to create an animal that hatches from an egg using scraps of construction paper, glue, and crayons.

- Have children use markers to draw "stitches" around the edges of the squares and write the name of the animal. Arrange squares to make a quilt entitled "What hatches from an egg?"

© Macmillan/McGraw-Hill

F is for Food Chains

Music

Food Chain Song

WHOLE CLASS

Objective: Identify a food chain.

Science Inquiry Skills: communicate, make a model, put things in order

Resources: Food Chain Song p. 69, Chain Pattern Sheet p. 70

- Teach children the "Food Chain Song."

- Make a copy of the song sheet on yellow paper and cut it out in the shape of the Sun. Make copies of the strips for the chain model on the other color of paper.

- Have children cut out the strips and make a paper chain using tape or glue to hold each link together.

- Tape the chain to the Sun.

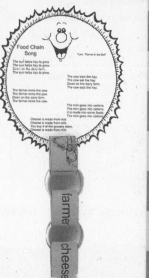

Reading

A Fishy Tale

WHOLE CLASS

Objective: Learn about a food chain.

Science Inquiry Skills: put things in order, make a model, communicate

Resources: A Fishy Tale p. 71

- Write "A Fishy Tale" on chart paper and read the story aloud to the class.

- Give children a copy of p. 71 and invite them to color the fish to correspond with the color words in the story.

- Help children cut out each fish and glue the fish to construction paper in the correct order, as shown.

- Encourage children to practice reading the story using the model they made.

Art

Bees and Bears

Materials

scissors, crayons, construction paper, glue

Objective: Understand that insects and animals can be a part of a food chain.

Science Inquiry Skills: communicate, put things in order

Resources: "The Bees and the Bears" Song p. 72, Activity Cards p. 73

- Teach children the song "The Bees and the Bears."

- Ask children to create a picture for each section of the activity cards. They can draw or use construction paper to create their pictures of a flower, a bee, a beehive, and a bear.

- Have children cut out the cards and play with them, putting them in correct sequence or making up their own story.

A flower grows in the garden.
A flower grows in the garden.
A flower grows in the garden
And makes the nectar sweet.

The bee sucks the nectar.
The bee sucks the nectar.
The bee sucks the nectar
From the pretty flower.

The bee flies to the beehive.
The bee flies to the beehive.
The bee flies to the beehive
Where the honey is made.

The bear climbs to the beehive.
The bear climbs to the beehive.
The bear climbs to the beehive
And takes the honey out.

Dramatic Play

Slippery Fish

Materials

9" x 12" light blue construction paper, craft sticks, scissors, crayons, tape

Objective: Make puppets and act out an imaginary food chain.

Science Inquiry Skills: communicate

Resources: Slippery Fish p. 74, Ocean Animals p. 75

- Chant the poem "Slippery Fish." Encourage children to add actions.

- Have children color all the characters from the poem, cut them out, and tape a craft stick to the back of each one.

- Fold the bottom of a piece of blue construction paper up one third of the way and draw waves with a blue marker. Cut the paper to look like a wave as well. Cut an opening at the fold and staple the sides.

- Invite children to stick their puppets through the hole, one at a time, as they chant the poem.

G is for Grow, Grow, Grow

Music

Germination

WHOLE CLASS

Materials

light blue paper, white paper, green paper, colored construction paper, lima beans, yarn, green pipe cleaners, green tissue

Objective: Learn the germination process.

Science Inquiry Skills: communicate, make a model

Resources: Germination Song p. 78

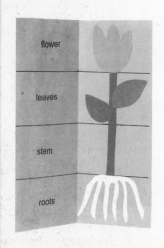

- Invite children to sing the "Germination Song." Engage children in a discussion about the song.

- Give each child a piece of light blue paper. Have children fold the paper in half the long way and hold horizontally. Help children draw four vertical lines and write the following labels, one in each section, along the bottom edge of the paper: *seed, roots, stem, leaves.*

- Help children glue lima bean seeds, yarn, pipe cleaners, and green tissue to illustrate each section as shown.

- Open the folded paper. Help children to draw four horizontal lines across the open paper and write the following labels along the left side: *flower, leaves, stem, roots.* Have children cut and glue the parts of a flower as shown.

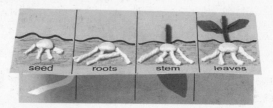

Art

What Makes a Plant Grow?

SMALL GROUP

Objective: Learn what a plant needs to grow.

Science Inquiry Skills: communicate

- Provide each child with a piece of square, white paper. Fold the paper in half where opposite corners meet in a triangle shape. Open and fold in the other direction. Unfold and then fold each corner into the middle.

- Help children label each flap: *sun, soil, water, air.* Have children illustrate each flap. Invite children to make a flower or plant underneath the flaps in the middle using construction paper or crayons.

Science

Watch a Plant Grow

SMALL GROUP

Materials

crayons, scissors, small plastic bags, $\frac{1}{2}$-cup measuring cup, potting soil, seeds (popcorn, beans, etc.)

Objective: Plant a seed and watch the sequence of its growth.

Science Inquiry Skills: observe

Resources: Watch Me Grow! Sheet p. 79

- Give each child a copy of the "Watch Me Grow!" sheet. Have children color and cut out the pot. Help children cut out the circle in the middle of the pot.

- Give each child a bag with $\frac{1}{2}$ cup potting soil in it. Have children plant 3 or 4 seeds in the soil. Have them add water until the soil is damp.

- Staple the bag to the back of the cut-out pot, making sure the bag remains slightly open. Staple these to a bulletin board and watch the parts of the plant grow and grow. Remind students to water their plants about twice a week.

Watch Me Grow!

Reading

What Would You Grow?

WHOLE CLASS

Materials

light blue paper cut into 3" x 8$\frac{1}{2}$" strips, construction paper (dark green, light green, yellow, orange)

Objective: Understand that we eat different parts of a plant.

Science Inquiry Skills: communicate, observe, compare

Resources: Tops and Bottoms by Janet Stevens, Graph Head pp. 76–77

- Read the story *Tops and Bottoms*. Discuss the 3 different plants the hare planted and why he planted them.

- Give each child a strip of light blue paper. Have children fold the bottom of the strip two-thirds of the way up and fold the top third down to make 3 sections.

- Have children create the crop they would grow, illustrating the roots, stem, and flower. Instruct children to make the roots in the bottom section, stem in the middle, and a flower in the top.

- Display the Graph Head on chart paper. Invite children to display their plants on the graph. Discuss which group has more and which group has fewer.

G is for Grow, Grow, Grow

 **is for Hibernation**

Music

Hibernating Habitats

SMALL GROUP

Materials

sentence strips, crayons, scissors, stapler, marker, chart paper

Objective: Introduce various places where animals hibernate.

Science Inquiry Skills: communicate, classify, put in order

Resources: H is for Hibernation Song p. 80, Animal Sheet p. 81, Habitat Sheet p. 82

■ Teach children the song "H is for Hibernation." Write the song on chart paper. As you sing, emphasize all the words that begin with the letter "H."

■ After children are familiar with the song, give each child copies of the animal and habitat sheets. Invite children to illustrate each habitat and color each animal. Have children cut out both sets of squares and put each set in order from 1–6.

■ Write "A ___ hibernates in ___." on a sentence strip for each child, and staple animal pictures after "A" and habitat pictures after "hibernates in."

■ Have children read the sentence strip by reading their first pictures, flipping them up and reading their second pictures, then continuing until all sentence versions are read.

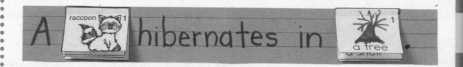

Science

Wake Up Bear

WHOLE CLASS

Materials

4" x 5" white paper, containers with hot and cold water, scissors, crayons, chart paper

Objective: Understand that weather affects hibernation.

Science Inquiry Skills: predict, investigate, draw a conclusion

Resources: Graph Head pp. 76–77

■ Display the Graph Head on chart paper. Have children draw a bear on a piece of white paper. Instruct them to fold the bear once vertically and once horizontally.

■ Encourage children to make predictions about which water, hot or cold, will "wake" the bear or make the bear unfold faster, and record their predictions on the graph.

■ Invite some children to drop their folded bears in hot water and others to drop their folded bears in cold water. Ask: **Which water caused the bears to "wake up," or unfold, faster?** Have children record their conclusions on the graph.

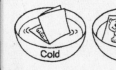

Reading

The "Bear" Facts

SMALL GROUP

Objective: Understand the cycle of hibernation.

Science Inquiry Skills: put in order, communicate

Resources: Bear Facts Song p. 83, Bear Pattern Sheet p. 84

- Copy the song "Bear Facts" on chart paper and teach children the song. Emphasize the different seasons for each verse.

- Make copies of the "Bear Facts" song and cut out each verse. Prepare a book for each child by folding white paper in half and gluing a verse on each page. Give each child a copy of the "Bear Facts" book to illustrate. Have children make a collage on each page to create a tree for each season.

- Have children color and cut out the bear pattern and glue it on a craft stick to use as a pointer as they read their book.

Materials

white paper, brown, green, and black construction paper, white paint, cotton swabs, stamp pads, pink tissue paper, scissors, crayons, glue, craft sticks, chart paper

---

Math

Hibernate or Not

WHOLE CLASS

Objective: Identify animals that hibernate and animals that do not hibernate.

Science Inquiry Skills: classify

Resources: Animal Pictures p. 85

- Invite children to brainstorm animals that hibernate and animals that do not. Record their ideas. Engage children in a discussion about animals that live through the winter in various settings.

- Have children draw an animal on paper. Write "Animals That Hibernate" and "Animals That Do Not Hibernate" on chart paper. Invite children to tape their drawings under the correct category and discuss.

- Give children the animal pictures to color and cut out. Then have children sort them according to hibernating (raccoon, turtle, bat, snake, mouse) and non-hibernating (monkey, cow, bird). Instruct children to fold a piece of construction paper in half horizontally, cut the top layer down the middle, and label each flap to match the chart. Then, have children glue pictures underneath the flap with the correct label.

Materials

construction paper, scissors, crayons, glue, chart paper, tape

I is for Insects

Science

Position Chart

WHOLE CLASS

Objective: Introduce insects and where they can be found.

Science Inquiry Skills: communicate, observe

Resources: Insects Sheet p. 86

- Make a position chart as shown. Copy the insects on cardstock and color and laminate them. Stick fastener material to the back of each insect.

- Label the chart: "on the leaves," "above the flower," "in the grass," and "under the ground." Place a piece of fastener material in each position on the chart.

- Have children place insects in the correct position on the chart.

Materials
14" x 22" poster board, colored construction paper, glue, hook-and-loop fastener material, cardstock

Math

Baffling Bugs

SMALL GROUP

Objective: Use logical thinking skills to listen and follow directions.

Science Inquiry Skills: communicate, put in order, draw a conclusion

Resources: Directions Sheet p. 87, Logic Line-Up Cards p. 88

- Give each child a copy of the Logic Line-Up Cards to color and cut out. You can also enlarge one set, color, and laminate them for children to hold in a whole group setting.

- Read the line-up directions one at a time. Have children line their cards up in front of them.

- In a whole group setting, invite children to give suggestions to the children holding the cards on which position they should move in order to follow the directions.

Materials
crayons, scissors

ladybug

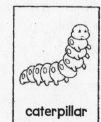
caterpillar

bee

grasshopper

Art — Incredible Quilt

SMALL GROUP

Objective: Sequence how a bee makes honey.

Science Inquiry Skills: communicate, put in order

- Have children cut a flower shape out of a small paper plate and color the center yellow. Have each child glue the flower and two green leaf shapes on each side of the flower on the blue square.

- Cut out an oval shape from black construction paper and have children glue it on the orange square. Invite children to use materials to decorate it to look like a bee, as shown.

- Have children brush a piece of bubble wrap with orange paint and have them press a piece of white drawing paper on it. Repeat with yellow paint and the same piece of paper. After the paint has dried, cut out a beehive shape, mount it on the green square, and have children stamp or draw 5 bees around the hive. Lay the quilt squares out in the following pattern: flower, bee, hive, etc.

Materials

squares of blue, orange, and green paper; small paper plates; yellow and black construction paper; tissue paper; orange and yellow tempera paint; bubble package wrap; bee stamps or stickers; drawing paper; markers; scissors; glue

---

Math — Irresistible Ladybugs

SMALL GROUP

Objective: Graph information using lady-bugs made with glyphs.

Science Inquiry Skills: compare, communicate

Resources: Glyph Direction Sheet p. 89

- Give each child a white square of paper and a black marker. Place red, yellow, and orange circles; small, rectangle-shaped strips of black paper; and cut-outs of green leaves in a pile.

- Read the glyph directions one at a time, allowing children to make a choice and create a ladybug according to their preferences. The directions can be reproduced on cards for children that need visual clues.

- After each child's picture is complete, sort and graph the answers to each question with the pictures. Line up how many orange, red, and yellow ladybugs were made, how many curly or straight antennas, etc.

- As an extension, create a skip-counting chart with the ladybugs, counting by 2s (antennas).

Materials

8" x 8" squares of white construction paper; 4" red, yellow, and orange paper circles; black and green construction paper; black markers

© Macmillan/McGraw-Hill

 # J is for Jumping Critters

Math — Jump Up

INDIVIDUAL

Materials
chart paper, construction paper, crayons, glue, scissors

Objective: Identify animals that jump and animals that do not jump.

Science Inquiry Skills: classify, compare

Resources: Animal Sorting Cards p. 90

- Encourage children to discuss animals and how they move. Have children brainstorm animals that jump and do not jump. Record their ideas on chart paper.

- Give each child a copy of the Animal Sorting Cards. Have children color and cut out the animals and sort them on a piece of paper with the left side labeled "Animals that can jump" and the right side labeled "Animals that cannot jump."

- Invite children to glue the animals to the paper. Have them discuss their work with a classmate.

Reading — Jumping Frog

SMALL GROUP

Materials
light blue construction paper, plastic or paper frog, craft sticks, crayons, stapler

Objective: Read position words.

Science Inquiry Skills: communicate

Resources: *Jump, Frog, Jump* by Robert Kalan, Sentence Strips Sheet p. 91

- Read the story *Jump, Frog, Jump*. Discuss places that the frog jumped on, by, into, etc.

- Have children draw a picture on light blue paper that includes a tree, flowers, a pond, and a large rock.

- Have children attach a paper or plastic frog to a craft stick. Help children cut out the sentence strips and staple them on the bottom left side of their drawings.

- Encourage children to read the strips and use the frog to show the jumping action.

The frog jumps on the rock.

_ jumps into the water.

Math

Just Measure

SMALL GROUP

Objective: Measure a frog picture using non-standard units of measurement.

Science Inquiry Skills: compare, measure

Resources: Just Measure Sheet p. 92

- Provide children with various items to measure the picture of the frog on the measuring sheet. Remind them to measure from top to bottom.

- Record all the results on chart paper. Encourage children to compare their results with a classmate's.

Materials
connecting cubes, toothpicks, beans, counting bears, linking loops, chart paper, pencils

Reading

Jumping Bunny

INDIVIDUAL

Objective: Read position words.

Science Inquiry Skills: communicate

Resources: Jump, Bunny, Jump! Story p. 93, Paper Bunny p. 94

- Make 4-page horizontal books for each child by folding and stapling two pieces of white paper together. Make copies of "Jump, Bunny, Jump!" cut out each sentence, and glue one sentence to the bottom of each page.

- Have children illustrate each page of the story with construction paper or crayons. On the last page, attach a small plastic bag.

- Help children cut out the paper bunny and tape one end of the yarn to the bunny and the other end to the first page of the book. Children can store the bunny in the plastic bag.

- Encourage children to make the bunny hop through the book as they read the story.

Materials
white paper, yarn, small plastic bag, construction paper, crayons, tape, glue

 **is for Kids Like to Touch**

Science — Kids Touch

WHOLE CLASS

Objective: Classify classroom objects according to how they feel.

Science Inquiry Skills: observe, compare, classify

hard	soft

rough	smooth

- Lay two jump ropes or long pieces of yarn across each other on the classroom floor to form four quadrants.

- Place a hard item in the first quadrant, a soft item in the second, a rough item in the third, and a smooth item in the fourth. Place a word label in each section.

- Invite children to look around the room for hard, soft, rough, and smooth items to add to the collection. You can also have children explore outside to find items.

- Make lists on chart paper of the items that children find.

Materials
jump ropes or long pieces of yarn, index cards, markers, chart paper, classroom items

Science — Kid Gloves

PAIRS

Objective: Understand that touching items gives us information.

Science Inquiry Skills: compare, communicate

- Provide a set of two objects. Put one object inside a box with some different objects and keep the other object outside of the box.

- Have children work with a partner. Have one child show the object to his or her partner. Without looking, the partner needs to find the same object in the box using their sense of touch.

- Invite children to try the activity again while wearing gloves. Ask: **Was it easy or difficult to identify the objects while wearing gloves? Why?**

Materials
gardening gloves, boxes, a variety of objects – 2 blocks, 2 cotton balls, 2 fabric pieces, 2 pine cones, 2 balls, 2 rocks, 2 sponges, 2 craft sticks, 2 marshmallows

© Macmillan/McGraw-Hill

Kid Blindfolds

SMALL GROUP

Materials
scarf; small paper clips; small container; rice; socks; plastic letters, numbers, or shapes

Objective: Use our sense of touch to find and count objects.

Science Inquiry Skills: predict, communicate, observe

- Invite children to take turns wearing the scarf as a blindfold or just cover their eyes. Have them reach in a small container filled with rice and small paper clips. Ask: **How many paper clips can you pull out in 30 seconds?**

- As an extension, put several plastic letters, numbers, or shapes in a sock and have children reach in and try to identify them.

Kid Fingerprints

SMALL GROUP

Materials
stamp pad, magnifying lenses, markers

Objective: Learn that fingerprints are different.

Science Inquiry Skills: observe, compare, draw a conclusion

Resources: Fingerprint Sheet p. 95

- Give each child a fingerprint sheet. Have children put a fingertip on a stamp pad and make a fingerprint in the box on the activity sheet.

- Invite children to use magnifying lenses to study the fingerprints and determine which type of fingerprint they each have.

- Encourage children to make more fingerprints at the bottom of the paper and transform the fingerprints into pictures.

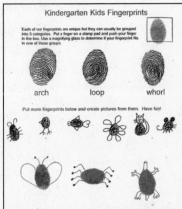

Kindergarten Kids Fingerprints

arch   loop   whorl

Put more fingerprints below and create pictures from them. Have fun!

L is for Life Cycles

Reading
Butterfly Life Cycle

WHOLE CLASS

Objective: Learn the life cycle of a butterfly.

Science Inquiry Skills: communicate, observe, make a model

Resources: *The Very Hungry Caterpillar* by Eric Carle

Note: Check children's allergies before you begin.

- Read *The Very Hungry Caterpillar* and discuss the life cycle of the butterfly.

- Give each child a strip of construction paper and have them fold it in half and then in half again, making four rectangles. Have children cut out and glue a large leaf shape on each rectangle. Below each leaf, help children label the different stages of the life cycle of a butterfly: egg, caterpillar, chrysalis, butterfly.

- Have children glue materials on the leaves as shown to illustrate the life cycle.

Materials
strips of 18"x 3" construction paper, green construction paper, tiny pom-poms, colored marshmallows, peanut shells, pipe cleaners, colored tissue, glue, scissors

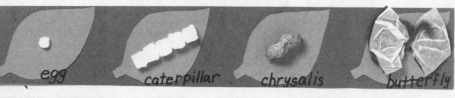

egg   caterpillar   chrysalis   butterfly

Math
Frog Life Cycle Quilt

SMALL GROUP

Objective: Learn the life cycle of a frog.

Science Inquiry Skills: observe, make a model, put things in order

- On the yellow square, have children stick 10 reinforcements in a cluster to represent frog eggs. Then help children cut 3 tadpole shapes out of green paper and glue them on the dark blue square.

- Have children trace their four fingers spread apart with the thumb tucked under on green paper. Have them cut out their tracing and glue it upside down on the light blue square. Then, have children design their frogs using craft eyes, markers, and stickers.

- Alternate the eggs, tadpoles, and frogs in a pattern on butcher paper. Use this quilt to help children skip-count by 3s.

Materials
square pieces of yellow, dark blue, light blue, and green construction paper; ring reinforcements; bug stickers; scissors; glue; craft eyes; markers

Chicken Life Cycle

SMALL GROUP

Materials
crayons, scissors

Objective: Learn the life cycle of a bird.

Science Inquiry Skills: observe, put things in order, make a model

Resources: Egg Activity Sheet p. 96

■ Have children illustrate the life cycle of a chicken by drawing pictures using each egg shape on the Egg Activity Sheet.

■ In the first box, have children draw a hen laying an egg on her nest. In the second box, a cracking egg should be drawn. In the third box, children should draw a chick. In the last box, children should draw a hen.

■ When the drawings are complete, the pictures can be cut apart for children to put in order. Ask: **Which came first, the hen or the egg?**

 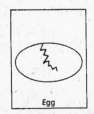  

Hen   Egg   Chick   Hen

Life Cycle of You!

INDIVIDUAL

Materials
scissors, crayons, tape, white paper, construction paper

Objective: Learn the life cycle of a human.

Science Inquiry Skills: observe, compare, put things in order

■ Give each child a piece of white paper divided into four quadrants. Have children bring in baby pictures and recent pictures of themselves to mount in the top two quadrants of the paper.

■ Have children draw what they think they will look like when they are teenagers and adults in the bottom two quadrants. Children can also draw all four pictures.

■ Help children cut apart the quadrants and glue them in order on a 6" x 18" strip of construction paper.

M is for Magnets and Much More

Magnets

SMALL GROUP

Objective: Identify objects that are attracted to magnets.

Science Inquiry Skills: predict, classify, investigate

Resources: Magnet Sheet p. 97

- Show children the collection of objects and have them predict which ones might be attracted by the magnet.

- Invite children to use magnets to test if they will attract the objects.

- Have children circle "yes" or "no" on the activity sheet to record the results.

Materials

magnets, nail, pencil, safety pin, small scissors, eraser, paper clip, box of crayons

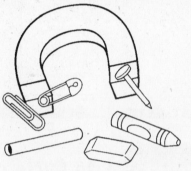

Math

More or Less

PAIRS

Objective: Determine which items in a group weigh more.

Science Inquiry Skills: predict, measure, communicate, compare

Resources: More or Less Sheet p. 97

- Encourage children to predict which item in each group will weigh more.

- Invite children to work with a partner to test the two items on the balance and circle the item that weighs more on the recording sheet. Have children continue until all sets of items have been tested.

- Invite groups to compare their results.

Materials

balance, clay ball, crayons, large eraser, wooden block, small scissors

© Macmillan/McGraw-Hill

Social Studies Magnifying Lens

SMALL GROUP

Materials
magnifying lenses,
pennies, paper, pencils

Objective: Use a magnifying lens to examine objects closely.

Science Inquiry Skills: observe, communicate

Resources: Magnifying Lens Sheet p. 98

- Invite children to use magnifying lenses to look closely at the pictures on both sides of a penny and determine the year it was minted.

- Have children write the year of the penny on the recording sheet. Encourage children to draw a picture of what they saw on each side of the penny.

- Then invite children to use a magnifying lens to examine the dinosaur pictures.

Writing Mirrors

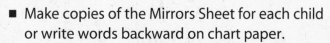
SMALL GROUP

Materials
small mirrors (that have
no edge piece around
them), paper

Objective: Use mirrors to determine words that are written backward and letters that are half-printed.

Science Inquiry Skills: observe, communicate

Resources: Mirrors Sheet p. 98

- Make copies of the Mirrors Sheet for each child or write words backward on chart paper.

- Invite children to place a mirror next to the words to determine what words are written backward. Then have children write the words correctly on a separate piece of paper.

- Encourage children to use the mirrors to identify the letters that are half-written. Have them complete the letters and practice writing the whole letter.

M is for Magnets and Much More

N is for Nature

Writing — Nature Story

WHOLE CLASS

Objective: Understand how a tree changes during the four seasons.

Science Inquiry Skills: communicate

Resources: Tree Sheet p. 99

- Put together "Nature's Story" books for each child by using 1 full tree and 3 treetop patterns stapled together with a cover as shown. Invite children to color the trunk and branches of each tree.

- Encourage children to change the top part of the tree according to each season. For fall, have children make yellow, red, and orange fingerprints to represent leaves. For winter, children can use cotton swabs to paint white dots on the branches to represent snowflakes. For spring, have children make flower blossoms out of yellow tissue paper. For summer, children can use green tissue paper and orange sticky dots to represent leaves and fruit.

- Help children write the name of the season or a sentence about what they like to do in each season on the corresponding pages.

Materials

colored tissue paper; white paint; cotton swabs; orange sticky dots; orange, red, and yellow ink pads; crayons; markers

Social Studies — Nature and Me

SMALL GROUP

Objective: Understand how to dress for a particular season.

Science Inquiry Skills: communicate, compare

Resources: Paper Doll Pattern p. 100

- Have children brainstorm what type of clothing is worn in each season. Provide each child with a copy of a paper doll. Invite children to create clothing and dress the paper doll according to their favorite season.

- Ask children to identify their favorite season and display their finished dolls on chart paper.

Materials

scraps of construction paper, crayons, markers, glue, chart paper

Nature's Weather Song

 WHOLE CLASS

Materials
five 8½" x 11" pieces of white cardstock, markers, yarn, chart paper

Objective: Recognize weather symbols and learn how to spell different types of weather.

Science Inquiry Skills: communicate, compare

Resources: Weather Song p. 101

- Make five weather signs by drawing pictures of different types of weather and writing labels on white cardstock. Fasten yarn to the weather signs so that they can be worn around a child's neck. Invite five volunteers to wear the weather signs.

- Teach children the "Weather Song." As children sing, have the child wearing the corresponding weather sign step forward so children can practice identifying the weather words. Repeat the song until all children have had a turn to wear a weather sign.

- After singing the song, ask children to name their favorite kind of weather. Graph the results on chart paper.

  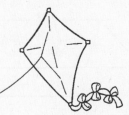 

Nature's Weather

SMALL GROUP

Materials
white drawing paper, crayons, markers, scissors

Objective: Understand that different activities take place in different types of weather.

Science Inquiry Skills: communicate

- Help children fold a piece of paper in half the long way and then fold in thirds. Three boxes should be visible.

- Invite children to select their three favorite types of weather. Help children to write the name of the type of weather at the top of each box.

- Cut along each fold to create three flaps. Have children lift each flap and draw a picture of what they like to do in each type of weather.

O is for Opposites

Science

Sink or Float

Objective: Test various items to determine if they sink or float.

Science Inquiry Skills: predict, observe, classify

Resources: Sink or Float Recording Sheet p. 102, Sink or Float Sorting Mat p. 103

- Lay out the items next to a tub of water or a sink full of water. Have children predict if each item will sink or float.

- Invite children to test the items. Have children circle "sink" or "float" on their recording sheet.

- After all children have done the experiment and completed their recording sheet, have them cut out the pictures and display the results on the Sink or Float Sorting Mat.

Materials

tub of water, key, balloon, banana, small rock, pencil, straw, crayon, toothbrush, coin

Math

Living and Nonliving

Objective: Determine if objects are living or nonliving.

Science Inquiry Skills: communicate, classify

Resources: Living and Nonliving Sorting Cards pp. 104-105

- Make a sorting mat by drawing two large circles next to each other on a piece of paper. Place the sorting mat and the sorting cards at a table.

- Invite children to work in pairs to discuss if the pictures show living or nonliving things.

- Have children place the cards on the correct side of the sorting mat.

Materials

scissors, 11" x 17" paper, marker

Living

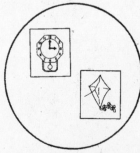

Nonliving

Hot and Cold Foods

Materials
file folders, crayons, scissors, paper, glue

Objective: Determine if certain food items are usually served hot or cold.

Science Inquiry Skills: classify, communicate, compare

Resources: Food Pictures Sheets pp. 106–107

- Make a "Hot and Cold Foods" game board using the inside of a file folder. Cut and glue a white piece of paper on each side of the inside of the folder. Label the left side "Food Served Hot" and the right side "Food Served Cold." Cut out the food pictures and have children color them.

- Encourage children to work in pairs to discuss each food item and determine if the food is usually served hot or cold.

- Have children put the pictures on the correct sides inside the folder.

Healthy and Unhealthy

Materials
sentence strips, markers, stapler, scissors

Objective: Role-play healthy and unhealthy habits.

Science Inquiry Skills: communicate, observe

Resources: Healthy and Unhealthy Habits Cards pp. 108–109

- Help children color and cut out the Healthy and Unhealthy Habits Cards and put them in a pile, picture-side down. Have volunteers pick a card and act it out. Encourage children to discuss the activities and determine whether they represent healthy or unhealthy habits.

- Sort the pictures into two piles. Ask children to identify and draw more healthy and unhealthy habits. Help children cut out and sort their pictures.

- Make a sentence strip for each child with the following text: "_____ is healthy but _____ is unhealthy." Staple the habit pictures in the appropriate places on the sentence strips to create mini-flipbooks as shown.

 is healthy but is unhealthy.

P is for Paper Fun

Math | Paper Planes

Materials
stickers, markers, tape measure or blocks

Objective: Measure how far a paper airplane can fly.

Science Inquiry Skills: predict, communicate, measure

Resources: Graph Head pp. 110–111, Airplane Sheet p. 112

- Provide each child with a copy of the paper airplane. Have children fold along the A line first, then the B lines and the C lines.

- Establish a means of measuring the distance the paper airplanes fly. Line up blocks or lay out a tape measure marking every 10 blocks or 10 feet.

- Display the Graph Head at the top of a piece of chart paper. Have children predict how far they think their plane will soar. Record their predictions by writing their names on the left side of the graph.

- Have children start at the same point and invite them to fly their airplanes. Have children fly their airplanes a few times and record the farthest flight on the right side of the graph. Discuss different ways that would help their plane fly farther.

Math | Powerful Paper

Materials
paper, counting bears, Water Table or container of water, blocks, paper cups, pennies, chart paper

Objective: Understand that papers have different strengths, which can be affected by folding.

Science Inquiry Skills: predict, compare, infer, communicate

Resources: Graph Head pp. 110–111

- Invite children to make a boat out of a piece of paper. Have children predict how many counting bears their boat will hold before it sinks. Display the Graph Head at the top of a piece of chart paper. Record children's predictions.

- Have children put their boat in water and count how many bears the boat actually holds until it sinks. Record results on the graph.

- Next build a bridge with a piece of paper and some blocks. Place a paper cup on the bridge and have children predict how many pennies the cup will hold before the bridge collapses. Accordion-fold a piece of paper and place it on the blocks. Ask children if the paper cup can hold more pennies on the folded paper or on the flat paper.

Paper Predictions

Materials
chart paper, different types of paper, eyedroppers, container of water

Objective: Learn what kinds of paper can and cannot absorb water.

Science Inquiry Skills: observe, predict, draw a conclusion

Resources: Paper Prediction Recording Sheet p. 113

- Make and display a chart of different types of paper as shown. Have children discuss the types of paper they think will absorb water and the ones they think will not. Encourage them to explain why.

- Place small pieces of different types of paper at a table. Give copies of the recording sheet to children and invite them to choose and write down four types of paper. Have children record their predictions by drawing a happy face if they think the paper will absorb water and a sad face if they think the paper will not.

- Invite children to do the experiment by dropping water on each type of paper with an eyedropper. Have children record their findings in the "Conclusion" column on the recording sheet.

People Popping

Materials
flat, low container; construction paper; copy paper; scissors

Objective: Understand that not all paper is water resistant.

Science Inquiry Skills: observe, predict, compare

Resources: People Poppers p. 114

- Help children cut out their people poppers, one on construction paper and one on copy paper. Have children fold the people's arms, legs, and head into the middle.

- Ask children if there is a difference in the two people they cut out. Have children predict what each paper person will do when placed in water.

- Invite children to place both of their paper people in the water at the same time and observe what happens. Invite children to share their results.

Q is for Quiet or Not

Music   Quite a Sound!

WHOLE CLASS

Objective: Learn about sounds in the environment.

Science Inquiry Skills: communicate

Resources: Oh, Can You Hear? Song p. 115

- Teach children the song "Oh, Can You Hear?"

- Ask children to create new verses to the song about sounds in their environment, such as about a baby crying, a bird singing, a cat meowing.

- Write children's new verses on sentence strips and place them in a pocket chart. Invite children to illustrate their new song.

Materials
pocket chart, sentence strips, drawing paper, crayons

Reading   Let's Listen

WHOLE CLASS

Objective: Identify sounds heard in school and in the neighborhood.

Science Inquiry Skills: observe, classify, communicate

Resources: The Listening Walk by Paul Showers

- Read the book *The Listening Walk*. Ask children to identify the sounds the little girl in the story hears in her neighborhood.

- Take children on a walk around the school. Provide paper, clipboards, and pencils and encourage children to draw or write sounds they hear. Ask children to think about whether the sounds are human-made or nature-made, for example, a car motor or leaves falling.

- Back in the classroom, make a chart with the titles "Human-made" and "Nature-made." As children identify sounds they heard, ask them to determine which category the sound fits under. Tally the results.

Materials
paper, pencils, clipboards, chart paper, marker

Human-made	Nature-made
car horn	bird
radio	falling leaves

Quick, Can You Pick?

PAIRS

Materials
pennies, toothpicks,
rice, sand, buttons,
pebbles, paper clips,
marbles, film canisters
or plastic eggs, stickers

Objective: Identify two canisters that contain the same objects by using the sense of hearing.

Science Inquiry Skills: compare, investigate, draw a conclusion

- Make two containers of each object. Be sure that children cannot see through the containers. Display the containers on a table with stickers on the bottom to indicate matches.

- Have children work at the Science Center with a partner as they shake each container to match the ones that have the same items by using their sense of hearing.

Writing

Quiet, Listen, and Guess

SMALL GROUP

Materials
5 small opaque containers with lids, small toy car, large button, ball, connecting cube, large safety pin, index cards, markers, paper, scissors, glue

Objective: Use the sense of hearing to identify an object.

Science Inquiry Skills: observe, predict, communicate

Resources: Listen Book p.116

- Seal the five objects in separate containers and label the containers 1–5. Illustrate each object on an index card and lay out the cards to show children the contents of each container.

- Invite children to shake the containers and lay them by the pictures that they think match the contents.

- Help children fold a piece of paper so that the two sides meet in the center as shown. Write "In the box is a …" on the flaps of the book. Have children cut out the boxes on page 116 and paste them inside the book. Then have them draw their guesses. After everyone has recorded their guesses, open the containers and check the results.

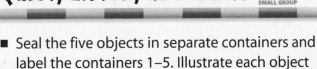

R is for Rocks

Science | Rock Scientists

SMALL GROUP

Objective: Learn what a petrologist is and discover unique things about rocks.

Science Inquiry Skills: communicate, observe, investigate

Resources: Petrologist Visor p. 117

- Explain to children that a petrologist is a scientist who studies rocks. Help children make their own visor to wear as they explore rocks. Form the headband of the visor using two lengths of yarn, or strips of paper stapled together.

- Provide children with tubs of water, paper towels, and toothbrushes to clean the rocks. Also provide magnifying lenses so children can closely study the rocks.

- Engage children in a discussion about their rocks and record the words these "petrologists" use to describe them on chart paper.

Materials
scissors, crayons, tubs of water, toothbrushes, paper towels, rocks, magnifying lenses, chart paper, markers, yarn or strips of paper

Math | Rock Sorting

WHOLE CLASS

Objective: Sort and classify rocks by their attributes, using a Venn diagram.

Science Inquiry Skills: observe, classify

Resources: Rock Sorting Labels p. 118

- Encourage children to look carefully and describe the rocks in the classroom rock collection. Have children sort the rocks in different categories.

- Introduce the words that are on the Rock Sorting Labels. Ask children to find a rock for each label as you talk about the words.

- Place two overlapping circles of string on the floor. Invite children to pick two labels to place in the circles. Have children sort the rocks accordingly. Ask: **What should we do with a rock that is both small and bumpy? Where does the rock go?**

Materials
rocks, string or yarn, scissors

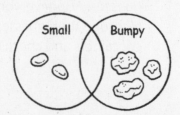

Rock Painting

SMALL GROUP

Materials
several different rocks, drawing paper, paint, paper towels, shallow containers

Objective: Learn that there are many different textures to rocks.

Science Inquiry Skills: classify, compare

- Make paint pads by folding paper towels in shallow containers and pouring small amounts of paint on them. Provide children with many different sizes and shapes of rocks with various outside textures.

- Have children dip rocks in paint pads and press them onto their drawing paper. Encourage children to see how many different texture prints they can create.

- Show children another fun way to utilize the paint and rocks. Encourage them to select one rock they would like to paint to use as a "pet."

Cooking

Rock Recipes

WHOLE CLASS

Materials
ingredients needed for the recipes (refer to recipe sheet), mixing bowl, measuring utensils, spoons

Objective: Make edible "rocks."

Science Inquiry Skills: communicate, observe, compare

Resources: Rock Recipes p. 119

Note: Check children's allergies before you begin.

- Make one recipe at a time with children. Allow children to sample both recipes and discuss the differences about how they taste and feel.

- Help children graph which recipe they liked best. Encourage them to share why they had a preference.

- You may want to send a copy of these recipes home for children to make with their families.

S is for Shadows

Reading

My Shadow and Me

WHOLE CLASS

Objective: Learn that shadows are different sizes at different times during the day.

Science Inquiry Skills: communicate, compare, observe, measure

Resources: *Bear Shadows* by Frank Asch, Paper Doll Pattern p. 100

- Read the book *Bear Shadows*. Discuss what causes shadows to appear and disappear, what happens to shadows when you move, and if shadows are always the same size. Invite children to color, cut out, and mount a paper doll on a craft stick to represent themselves.

- Take children outside in the morning and find a place in the ground to stick their paper doll. Use nonstandard units to measure the paper doll's shadow and record the length of each shadow. Repeat in the afternoon.

- Compare the differences between the shadows in the morning and the afternoon. Discuss with children why the shadow length was different.

Materials

white paper, crayons, scissors, glue, craft stick, nonstandard measuring units (straws, connecting cubes, blocks)

Art

Sun and Shadow Quilt

SMALL GROUP

Objective: Learn that shadows occur wherever a light source, such as the Sun, shines upon an opaque object.

Science Inquiry Skills: observe, communicate

- Take children outside to observe their shadows. Then invite children to draw themselves on manila paper with their body making a motion. Have children place a black piece of paper behind their drawing to represent their shadow. Help them cut out their picture along with the black paper. Have children mount their self portrait, with their shadow, on a light blue paper square.

- Encourage children to paint a sun on the orange paper square. When the suns are dry, have children cut them out, decorate them, and mount them on the dark blue paper. Display the squares in an A-B pattern: sun, shadow, sun, shadow.

Materials

dark blue, light blue, orange, manila, and black 8" x 8" squares of construction paper; yellow paint; glue; scissors; crayons

© Macmillan/McGraw-Hill

Shine Through

SMALL GROUP

Materials
aluminum foil, wax paper, paper towel, cardboard, clear plastic bag, flashlights

Objective: Learn that light can pass through certain materials.

Science Inquiry Skills: predict, communicate, investigate

Resources: Shine Through Recording Sheet p. 120

- Have children explore the different objects listed on the recording sheet to determine if light can or cannot shine through them. Have children predict which objects will allow light to pass through by circling "yes" or "no" in the prediction box.

- Invite children to experiment with each object by using a flashlight. Have children record their results by writing "yes" or "no" in the conclusion box. After all children have had a chance to experiment, discuss with them why light can or cannot pass through certain objects.

Science

Timely Shadows

WHOLE CLASS

Materials
large foam cups, wooden skewers, permanent markers

Objective: Learn that the shadows from the Sun help us to tell time.

Science Inquiry Skills: observe, compare, communicate

- Have children write their name on their foam cup. Bring children outside and invite them to place their cups upside down on the grass in a sunny spot.

- Help children push a skewer through the bottom of the cup and into the ground so that it will not move. Have children locate the skewer's shadow on the bottom of the cup and trace over the shadow with a marker. Encourage children to mark the skewer's shadow on the cup every hour to show how the Sun moves across the sky.

- Discuss with children that the position of the Sun can be used to tell time, as used on a sundial.

T is for Traveling

Social Studies — Texture Tryouts

SMALL GROUP

Objective: Understand that wheels roll on a variety of surfaces with different results.

Science Inquiry Skills: predict, investigate

Resources: Texture Tryouts Background p. 121, Texture Tryouts Story pp. 122–123

- Encourage children to name some smooth and rough surfaces. Discuss whether they think the type of surface affects how fast or slow a car can move.

- Have children predict if a plastic car will go fast, go slow, or not move at all on dry sand, wet sand, a plank of wood, a rug, a tile surface, a sidewalk, or a grassy area. Have them test their cars on all of the surfaces and record their results by circling "fast," "slow," or "no" on the story pages.

- Invite children to cut out their pages. Help children to staple them to the "Texture Tryouts" sheet as shown.

Materials
small plastic cars, wet and dry sand, wood plank, scissors, stapler

Art — Wheel Away

WHOLE CLASS

Objective: Understand that vehicles with wheels can move in various directions and over many surfaces.

Science Inquiry Skills: communicate, make a model

Resources: Wheel Away Story Sheet p. 124

- Discuss the directions that vehicles with wheels can move and the surfaces that vehicles can move on.

- Make 6-page horizontal booklets by gluing each phrase of the story on its own page as shown. Have children use construction paper and markers to illustrate their booklets to describe vehicle motion.

- Have children attach one pasta wheel to a piece of yarn and tape the yarn to the inside cover of their booklet. Encourage children to move the wheel along the various surfaces in a variety of motions.

Materials
wheel-shaped pasta, yarn, construction paper, markers, 6-page booklets

Reading

Transportation Trivia

WHOLE CLASS

Materials
chart paper, crayons,
markers, scissors

Objective: Listen to and solve transportation riddles.

Science Inquiry Skills: communicate, infer

Resources: Transportation Riddles Sheet p. 125, Transportation Cards p. 126

- Make a vertical booklet and glue each riddle on a page of the booklet as shown. Title the booklet "Can You Guess How People Move?"

- Read the riddles to children and have them answer each one with the type of vehicle described. Repeat the riddles and invite children to complete their booklets by illustrating the pages with the vehicles.

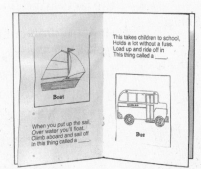

Music

Let's Go Traveling

WHOLE CLASS

Materials
3 sheets of white paper
per child, construction
paper, chart paper,
scissors, glue

Objective: Learn a song about various ways to travel.

Science Inquiry Skills: classify, communicate

Resources: Let's Go Traveling Song p. 127

- Teach children the song "Let's Go Traveling." Discuss with children where each of these vehicles travels.

- Write the words *air, land,* and *water* on chart paper. Invite children to name a variety of vehicles that move on air, land, and water and record their responses.

- Help children create a booklet by folding paper and cutting two lines 2 inches apart and 2 inches into the middle of each fold to form a section that can stand out. The booklet should include three pop-out sections.

- Have children create a cover and cut out construction paper vehicles for each section of their booklet. Have them label each page with *land, air,* or *water.*

U is for Universe

Science — The Sun

SMALL GROUP

Materials

yellow tissue paper, orange copy paper, white paper, scissors, glue-and-water mixture

Objective: Learn facts about how the Sun influences our daily lives.

Science Inquiry Skills: communicate

Resources: Sun Activity Sheet p. 128

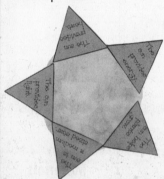

- Make copies of the Sun Activity Sheet on orange paper and give to children. Discuss with children information about the Sun, especially the facts listed on the five triangles.

- Help children cut out a circle from white paper and then the five orange triangles. Invite children to cover the circle with small squares of yellow tissue paper. Help children add a glue-and-water coating to the tissue.

- After the tissue is dry, have children glue the triangles around the circle to create a sun.

Science — Stars

WHOLE CLASS

Materials

star stickers, black paper, white crayons

Objective: Learn that people see patterns in stars and that these patterns are called constellations.

Science Inquiry Skills: observe, communicate, make a model

Resources: Constellation Sheet p. 129

- Show children the Constellation Sheet and discuss what constellations are.

- Provide each child with a Constellation Sheet, black paper, star stickers, and a white crayon. Have children create one of the six constellations as shown, connecting the stars with a white crayon.

- Encourage children to create and name a new constellation on a second sheet of black paper.

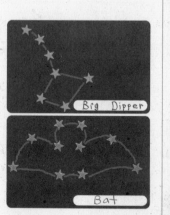

Reading

Planets

Materials
crayons, pencils, scissors, stapler, chart paper

Objective: Learn a poem about planets.

Science Inquiry Skills: compare, communicate

Resources: Planets Activity Booklet pp. 130, 131, 132

- Teach children the poem from the activity booklet and write it on chart paper or sentence strips.

- Help children make the activity booklet. Have them fill in the missing opposite words, color the pictures, then staple the booklet pages together.

- Encourage children to read the booklet to a classmate. Then have children take it home to share with a family member.

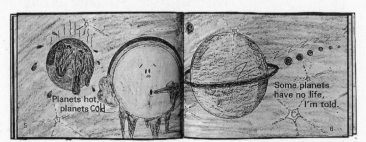

Planets hot, planets cold

Some planets have no life, I'm told.

Writing

Unique Shapes

SMALL GROUP

Materials
crayons, scissors, glue

Objective: Review and write words about basic shapes.

Science Inquiry Skills: observe, communicate, classify

Resources: Unique Shapes Booklet pp. 133, 134; Space Pictures p. 135

- Have children color the three space pictures and cut them out.

- Prepare Unique Shapes Booklets for each child as shown. Invite children to match the space pictures to the correct shape in their booklet.

- Have children write or dictate descriptive words about each object in space.

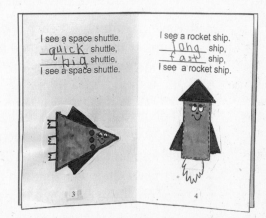

I see a space shuttle.
quick shuttle,
big shuttle,
I see a space shuttle.

I see a rocket ship.
long ship,
fast ship,
I see a rocket ship.

V is for Vision

Math — Eye Color Chart

WHOLE CLASS

Objective: Learn about vision and eye color.

Science Inquiry Skills: observe, investigate, communicate

Resources: "Did You Ever See?" Song p. 136; Eye Color Chart p. 137

- Teach the song "Did You Ever See?" to children and encourage them to act it out.

- Provide each child with an Eye Color Chart. Have children observe ten classmates' eyes and record the eye color.

- Help children write the names of the children they observed in the appropriate column on the Eye Color Chart.

Materials
crayons, pencils

Name
Look at 10 friends' eyes and record the colors. Start coloring at the bottom of each column and go up. Share your results with a partner.

Eye Color Chart

Austin		
Ofto		
Sara		Ella
Caitlin	Jeremy	Anna
Raya	Jared	Noah
brown	green	blue

Art — Optical Illusions

WHOLE CLASS

Objective: Explore and create optical illusions.

Science Inquiry Skills: observe, communicate

Resources: Optical Illusion Sheets pp. 138, 139

- Have children color the clown faces, the bird, and the nest on p. 138. Have them cut out the rectangles and fold each in the middle. Help children place a straw inside the fold and tape it down. Next, have children tape all of the sides of the paper.

- Have children twist each stick between the palms of their hands to see the clown face change and the bird go into the nest. Invite children to create their own illusions.

- Then invite children to look at the two pictures on p. 139 with their noses touching the paper. The bird will appear to go to the birdhouse and the duck will be in the water. Invite children to create their own optical illusion at the bottom of the page.

Materials
straws, crayons, scissors, tape

Science

Visual Memory

Materials
small classroom items, small blanket or piece of cloth

Objective: Observe and recall missing items.

Science Inquiry Skills: observe, communicate

- Encourage children to work in pairs. Have one child lay out objects on a table or tray while the other child studies them. Then have the child who laid out the objects cover them with a small blanket or piece of cloth while the other child tries to remember as many of the objects as he/she can. Have children switch roles and repeat.

- Then have one child remove an item while the partner closes his/her eyes. When the child opens his/her eyes, have him/her guess what is missing. Encourage children to take turns trying to identify the missing items.

Science

Vision in Light and Dark

Materials
3 pairs of dark socks (each a different color), 1 pair of white socks, small box or shoe box, beach towel

Objective: Learn that we need light to see.

Science Inquiry Skills: observe, communicate

- Have children work in pairs to match the socks in their box.

- Then have children go under a table covered with a beach towel, turn off the classroom lights, and have them try to match the socks again.

- Ask children if it was easier to match the socks with the lights on or off. Encourage children to explain why they think so.

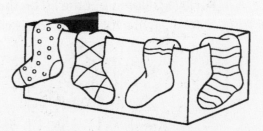

W is for Water

Reading — The World of Water

SMALL GROUP

Materials
crayons, scissors, staples or small rings

Objective: Learn the water cycle.

Science Inquiry Skills: communicate, observe

Resources: World of Water Book pp. 140, 141

- Discuss with children where rain comes from and where it goes when it falls. Show "The World of Water" pages to children and read the sequence of the water cycle.

- Give each child his/her own copy of the pages. Have children color the pictures and add details to the drawings. Help children bind the book on one side with staples or small rings.

- Read the book together again, flipping the pages to illustrate the water cycle.

The Sun heats the ocean.

2

Science — Water Wave

Materials
jars with lids, water, vegetable oil, blue food coloring, chart paper

WHOLE CLASS

Objective: Learn that liquids of different densities do not mix.

Science Inquiry Skills: compare, predict, communicate, observe

- Have children experiment with mixing liquids of different densities. Help children fill a jar $\frac{3}{4}$ of the way with water. Have children add a few drops of food coloring and then fill the jar the rest of the way with oil.

- Before you mix the ingredients, ask: **What do you think the water and oil will do? Will they mix and change color?** Record children's responses on chart paper.

- Have children put the lid on the jar and gently rock it back and forth. Invite volunteers to describe what happens.

Food Coloring

Oil

Math

Water Wonder

WHOLE CLASS

Materials

chart paper, large bowl of water, pennies, markers, dimes, nickels, quarters

Objective: Investigate how many coins it takes to make a bowl of water spill.

Science Inquiry Skills: predict, investigate, observe, compare

- Fill a bowl to the top with water. Have children predict how many pennies it will take to make the water spill over the top of the bowl and record their predictions on a chart titled "Will It Spill?"

- Invite children to slowly drop pennies into the water, one at a time. Have children count the pennies as they drop them into the bowl until the water spills over. Record results on the chart. Help children compare their predictions with the results.

- Encourage children to try this experiment using dimes, nickels, or quarters.

Will It Spill?	
Predictions	**Conclusions**
50	5
13	
5	
22	

Science

Water Wiggle

SMALL GROUP

Materials

wax paper, eyedropper, toothpicks, water, tape

Objective: Compare what happens to water on different surfaces.

Science Inquiry Skills: observe, compare

Resources: Water Wiggle Maze p. 142

- Have children place a drop of water at the start of their Water Wiggle Maze and try to drag it with a toothpick along the maze trail.

- Ask: **What do you think will happen to the drop of water when you try to drag it?**

- Then have children lay a piece of wax paper over the maze and tape it to the sheet. Invite children to repeat the activity, noting and comparing the results.

 **X** is for X-Ray

Math

"X-tra" Special Me!

SMALL GROUP

Objective: Understand that an X-ray is a picture of bones.

Science Inquiry Skills: make a model, observe, communicate

Resources: "X-tra" Special Me! Sheets pp. 143, 144

- If possible, show children some real X-rays. Invite them to identify the body parts. Explain to children that an X-ray is a special picture that doctors use to see teeth and bones inside their bodies.

- Provide children with copies of the "X-tra" Special Me! Sheets. Have children color them to look like themselves. Invite children to hold up their pictures, one on top of the other, to the light to see an X-ray of their whole body.

- Have children draw a shape for a head on white paper, cut it out, and glue it to a piece of black construction paper. Have children glue cotton swabs on the paper in a pattern to represent their bones.

Materials

black construction paper, cotton swabs, crayons, glue, white paper, scissors

Music

Hand X-Rays

SMALL GROUP

Objective: Learn that our hand is made up of many bones.

Science Inquiry Skills: make a model, communicate, observe

Resources: "Them Hand Bones" Song p. 145

- Discuss with children how many bones they think are in their hand. Teach children the song "Them Hand Bones."

- Invite children to trace their hand on a black piece of construction paper with white chalk or a white crayon.

- Help children cut straws in different lengths, 2 inch, 1 inch, and $\frac{1}{2}$ inch. Have children glue the straws inside their traced hands to resemble bones.

Materials

black construction paper, white chalk or crayon, white straws, scissors, glue

My Hand X-ray

© Macmillan/McGraw-Hill

Math

"X-tinct" Creatures

Materials
toothpicks, craft sticks, glue

Objective: Construct imaginary dinosaur skeletons.

Science Inquiry Skills: make a model, communicate

Resources: "X-amine X-tinct" Creatures Sheet p. 146

- Distribute the dinosaur sheet and invite children to build the skeleton of a dinosaur on it. Instruct students to use craft sticks and toothpicks for bones and glue them on the sheet.

- Encourage children to count how many bones they had to use to create such a large animal.

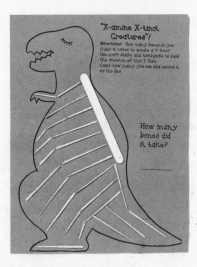

Science

Animal X-Rays

Materials
transparency sheets, 6" x 9" black construction paper, crayons, scissors, stapler

Objective: Match an animal X-ray to the correct animal.

Science Inquiry Skills: communicate, observe, compare

Resources: Animal Sheet p. 147, Animal X-Ray Sheet p. 148

- Photocopy pictures of the animals and have children color them and cut the boxes out. Photocopy the skeletons of the animals on transparency sheets. Have children cut them out. Help children lay the transparency of the animal skeleton on top of the correct animal drawing.

- Have children create a booklet entitled "My Animal X-Ray Book." Help children staple the animals and their skeletons to the inside of the booklet as shown.

Y is for Young Animals

Reading

Young Animal Names
WHOLE CLASS

Materials
sentence strips, crayons, scissors, glue

Objective: Match baby animal pictures to their names.

Science Inquiry Skills: classify, compare

Resources: Young Animals and Their Names Sheets pp. 149, 150

- Write the poem from "Young Animals and Their Names" on sentence strips to teach children about animal names.

- Give children copies of the Young Animals Sheet and have them color the pictures and cut them out. Make books and invite children to match the young animals to the correct sentence in their book. Then have children glue the young animals into their book. Reread the book together as a group.

Reading

Young and Old
WHOLE CLASS

Materials
crayons, pencil, chart paper

Objective: Discuss and draw activities that children are old enough or too young to do.

Science Inquiry Skills: communicate, compare

Resources: *Too Old for Naps* by Jane Yolen, Young and Old Sheet p. 151

- Read the story *Too Old for Naps*. Discuss activities children are old enough to do and activities they are too young to do. Write children's ideas on chart paper.

- Invite children to choose activities from the list and draw pictures of the activities on the Young and Old Sheet.

- Compile children's pages into a class book and read the book together as a group.

Who Am I?

WHOLE CLASS

Materials
paper, markers, crayons, scissors, tape

Objective: Identify animals based on "Yes" or "No" questions.

Science Inquiry Skills: communicate, predict

Resources: *Is Your Mama a Llama?* by Deborah Guarino

- Read the story *Is Your Mama a Llama?* Discuss with children the clues the baby llama used to identify the other baby animals.

- Draw and attach pictures of animals (monkey, turtle, skunk, lamb, frog, rabbit, bird, cat) to the backs of some children. Do not let children see which animal is on their back. Have children ask their classmates yes or no questions to identify the animal on their back.

Alike and Different

SMALL GROUP

Materials
white drawing paper, crayons, scissors

Objective: Identify differences and similarities between grown animals and baby animals.

Science Inquiry Skills: observe, compare, communicate

- Discuss the differences and similarities between grown animals and baby animals. Give children a piece of white paper, help them fold the paper in half horizontally, and cut three equal-sized flaps as shown.

- Encourage children to choose an animal, draw the adult animal on the top left flap, and draw the baby animal on the top right flap.

- Have children lift the left and right flaps and write or dictate something about the adult and baby animals. Under the middle flap, have children write or dictate something about how the two are alike.

Y is for Young Animals

Z is for Z-z-z at Night

Science

Diurnal or Nocturnal?

WHOLE CLASS

Materials
chart paper, markers, crayons, scissors

Objective: Understand the difference between diurnal and nocturnal.

Science Inquiry Skills: classify, infer, communicate

Resources: Animal Pictures p. 152

- Discuss the meaning of the words *diurnal* and *nocturnal*. Explain that a diurnal animal is one that is active mainly during the day and a nocturnal animal is one that is active during the night.

- Have children color and cut out the animals on the animal sheet. Make a class graph with the heads "Diurnal" and "Nocturnal." Invite children to select a picture to place under the appropriate heading on the graph.

- Encourage children to think about other animals that are diurnal or nocturnal. Have them draw pictures of these animals and add them to the graph.

Diurnal	Nocturnal

Music

Who's Awake?

SMALL GROUP

Materials
brass fasteners; star stickers; black, green, and brown construction paper; plastic wiggle eyes; scissors; crayons; glue

Objective: Recognize animals that are nocturnal.

Science Inquiry Skills: communicate

Resources: *Animals at Night* by Sharon Peters, "Nighttime Noises" Song p. 153

- Read *Animals at Night* and teach children the song "Nighttime Noises."

- Cut out tree trunks and treetops from colored construction paper. Have children glue a tree trunk to a black piece of construction paper. Help children cut a hole in the black paper at the top of the trunk.

- Cut out large circles from white paper and draw lines dividing each circle in fourths. On each section of the circle, have children draw an animal that is awake mainly at night. Have children attach the circle behind the tree with a brass fastener. Then have them glue the top half of the green treetop to the black paper, covering the hole as shown. Help children write "Who's awake tonight?" on the green treetop.

Reading

The Bear and the Moon

WHOLE CLASS

Objective: Identify different phases of the Moon.

Science Inquiry Skills: communicate, compare

Resources: *Moonbear* by Frank Asch, Moon and Bear Sheet p. 154

■ Read *Moonbear* by Frank Asch. Invite children to sponge-paint yellow stars randomly on a piece of black construction paper. Have children use torn pieces of colored construction paper to make a tree trunk and treetop.

■ Have children color, cut out, and glue the bear under the tree. Invite children to color the four Moon phases and cut out the strip of Moons. Help children cut two 3-inch long slits $2\frac{1}{2}$ inches apart at the top of the black construction paper to slide the phases of the Moon through as shown.

Materials

12" x 18" black construction paper; purple, green, and brown paper; yellow paint; star-shaped sponge; 3" x 10" strips of white drawing paper, crayons, scissors, glue

Math

Where Is the Bear?

SMALL GROUP

Objective: Practice using positional words.

Science Inquiry Skills: observe, communicate, put in order

Resources: *Moongame* by Frank Asch, Sentence Strips p. 155, Bear Patterns p. 156

■ Read *Moongame*, in which Moonbear plays hide-and-seek with the Moon.

■ Invite children to draw an outdoor scene on a piece of construction paper. Help children cut out sentence strips and staple them to the bottom of the drawing paper. Have children color the bear pattern and glue it to a craft stick. Have children glue a yellow circle at the top of the picture to represent the Moon.

■ Encourage children to read the strips and move the bear to the appropriate position.

Materials

construction paper, craft sticks, yellow circles, crayons, stapler, glue

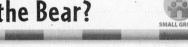
Bear and the Moon

How many ants

Prediction

Can it make

Prediction

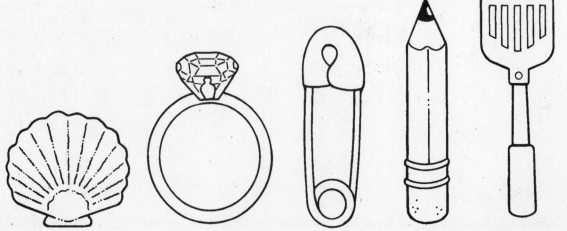

Use with **p. 2 Ant Story**
Use with **p. 4 Blowing Bubbles**

fit in your hand?

Conclusion

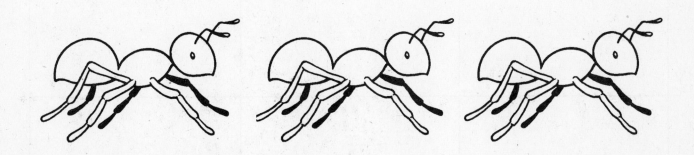

bubbles?

Conclusion

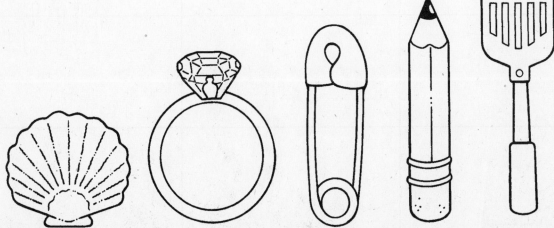

Name _____

Can it make bubbles?

Object	Prediction	Results

Use with **p. 5 Bubble Experiments**

Mixing Colors Song

(Tune: "If You're Happy and You Know It")

When you mix red and yellow, you get orange.
When you mix red and yellow, you get orange.
When you mix these two, you'll see
  a new color magically.
When you mix red and yellow, you get orange.

When you mix red and blue, you get purple.
When you mix red and blue, you get purple.
When you mix these two, you'll see
  a new color magically.
When you mix red and blue, you get purple.

When you mix yellow and blue, you get green.
When you mix yellow and blue, you get green.
When you mix these two, you'll see
  a new color magically.
When you mix yellow and blue, you get green.

Crayons for "Mixing Colors Song"

red

yellow

orange

red

blue

purple

yellow

blue

green

Use with **p. 6 Color Song**

Name _____

Color Mixing

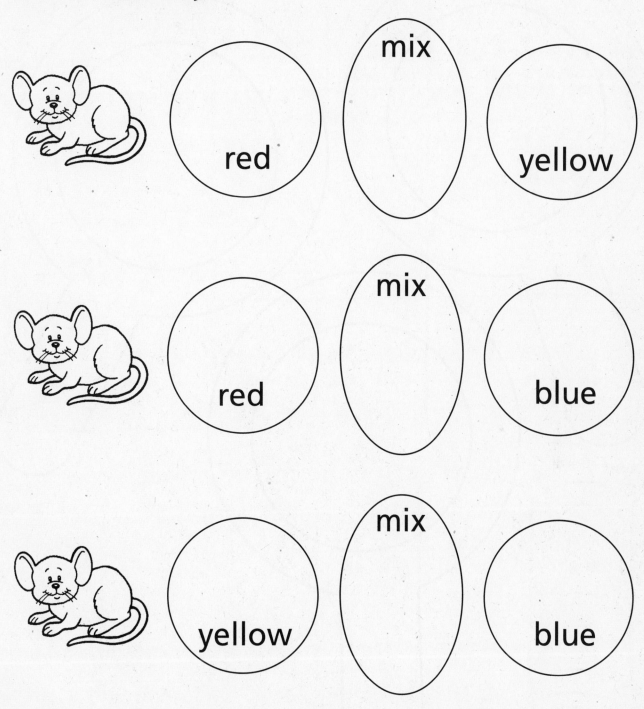

red   mix   yellow

red   mix   blue

yellow   mix   blue

Directions: Mix paints in the middle with craft sticks.
Use a new stick for each section.

Color Paddles

Use with **p. 6 Color Song**

Crayons for Color Quilt

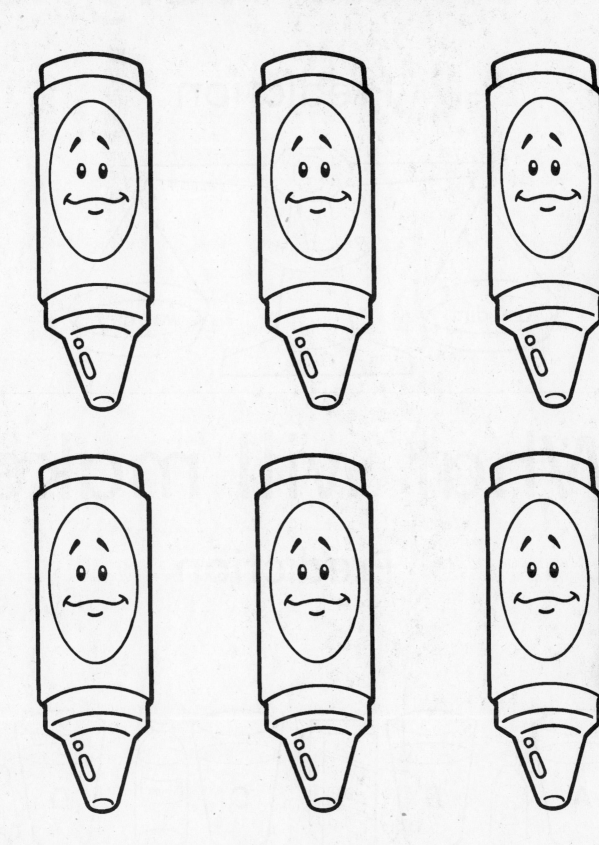

Which weighs

Prediction

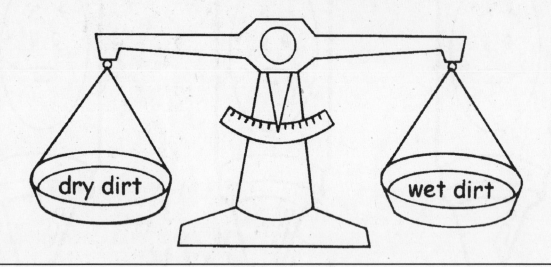

What will make

Prediction

Use with **p. 8 Weighing Dirt**
Use with **p. 11 The Ever-Floating Egg**

more?

Conclusion

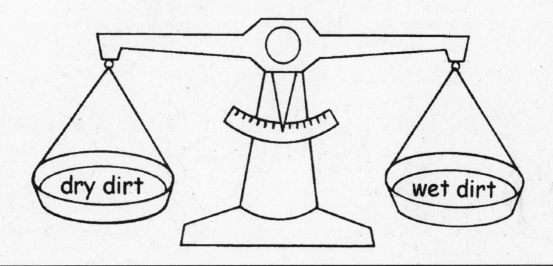

dry dirt   wet dirt

the egg float?

Conclusion

A   red

B   Corn Starch

C   vinegar

D   salt

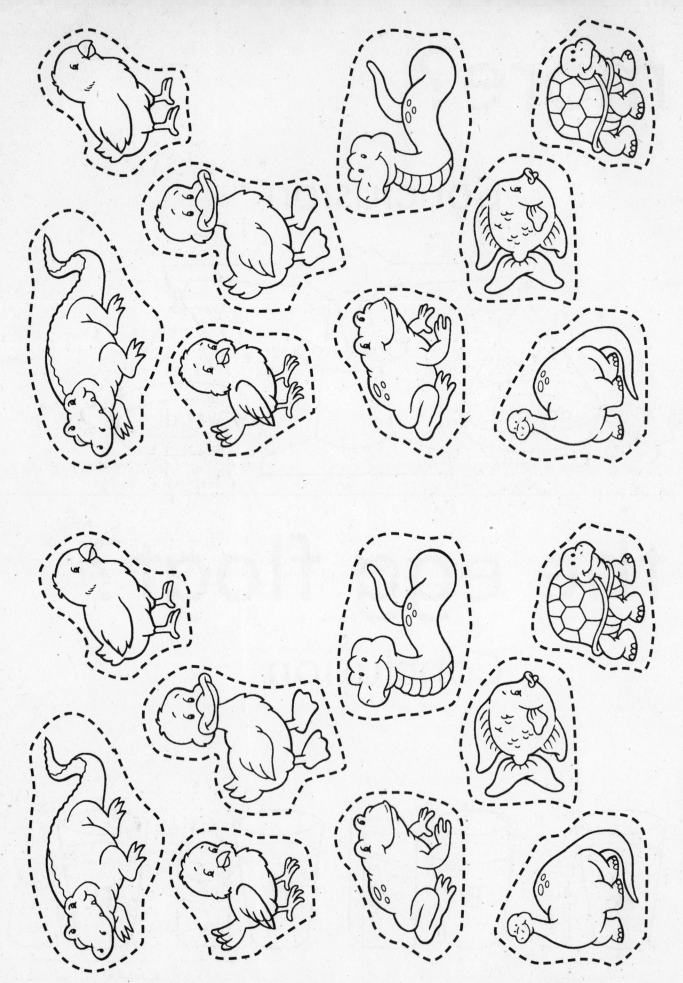

Use with p. 10 Match Book Riddles

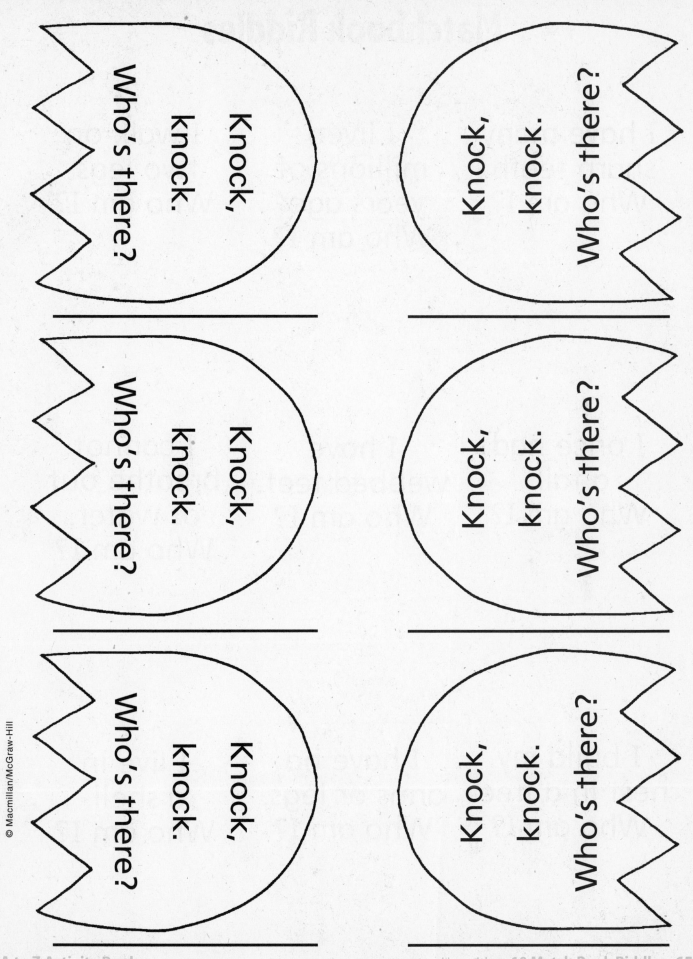

Knock,
knock.

Who's there?

Knock,
knock.

Who's there?

Knock,
knock.

Who's there?

Knock,
knock.

Who's there?

Knock,
knock.

Who's there?

Knock,
knock.

Who's there?

Matchbook Riddles

I have many
sharp teeth.
Who am I?

I lived
millions of
years ago.
Who am I?

I walk on
two legs.
Who am I?

I once had
a tail.
Who am I?

I have
webbed feet.
Who am I?

I cannot
breathe out
of water.
Who am I?

I build my
nest in a tree.
Who am I?

I have no
arms or legs.
Who am I?

I live in
a shell.
Who am I?

Use with p. 10 Match Book Riddles

An Extraordinary Egg

Dear Family,

   We have been learning about eggs and what hatches from them. Please read the enclosed book, *An Extraordinary Egg* by Leo Lionni, with your child. After reading the book, help your child find something he/she can put in the provided plastic egg. It can be a picture, a small toy, or something made from clay or wood. Be creative. Then help your child think of three clues about what is in the egg so that he/she can tell them to the class to help them guess the item.

   It is important that you return this extraordinary egg tomorrow as other children are anxiously waiting to have their turn to take the egg home and read the story with their family.

   We will be displaying all the items collected so we can see all the extraordinary things this egg has hatched!

   Thank you for being a part of our learning fun!

Write clues here:

1. _____

2. _____

3. _____

Name _____

The Ever-Floating Egg
Recording Sheet

1. Which solutions will make the egg float?
   Circle your guess. (Prediction)

2. Which solutions made the egg float?
   Circle your answer. (Conclusion)

Use with **p. 11 The Ever-Floating Egg**

Food Chain Song

(Tune: "Farmer in the Dell")

The Sun helps hay to grow.
The Sun helps hay to grow.
Down on the dairy farm,
The Sun helps hay to grow.

The cow eats the hay.
The cow eats the hay.
Down on the dairy farm,
The cow eats the hay.

The farmer milks the cow.
The farmer milks the cow.
Down on the dairy farm,
The farmer milks the cow.

The milk goes into cartons.
The milk goes into cartons.
It is made into some foods.
The milk goes into cartons.

Cheese is made from milk.
Cheese is made from milk.
You buy it at the grocery store.
Cheese is made from milk.

Food Chain Pattern

glue		hay
glue		cow
glue		farmer
glue		milk
glue		cheese

Use with **p. 12 Food Chain Song**

A Fishy Tale

Inside the blue fish, there is...

a purple fish.

And inside the purple fish, there is...

a yellow fish.

And inside the yellow fish, there is...

a green fish.

And inside the green fish, there is...

an orange fish.

And inside the orange fish, there is...

a red fish.

And inside the red fish, there is a hook.

1

2

3

4

5

6

© Macmillan/McGraw-Hill

The Bees and the Bears

(Tune: "The Bear Went Over the Mountain")

A flower grows in the garden.
A flower grows in the garden.
A flower grows in the garden
And makes the nectar sweet.

The bee sucks the nectar.
The bee sucks the nectar.
The bee sucks the nectar
From the pretty flower.

The bee flies to the beehive.
The bee flies to the beehive.
The bee flies to the beehive
Where the honey is made.

The bear climbs to the beehive.
The bear climbs to the beehive.
The bear climbs to the beehive
And takes the honey out.

Use with p. 13 Bees and Bears

A flower grows in the garden.
A flower grows in the garden.
A flower grows in the garden
And makes the nectar sweet.

The bee sucks the nectar.
The bee sucks the nectar.
The bee sucks the nectar
From the pretty flower.

The bee flies to the beehive.
The bee flies to the beehive.
The bee flies to the beehive
Where the honey is made.

The bear climbs to the beehive.
The bear climbs to the beehive.
The bear climbs to the beehive
And takes the honey out.

Slippery Fish

Slippery fish, slippery fish
Sliding through the water.
Slippery fish, slippery fish
    Gulp, gulp, gulp

Oh, no!
It's been eaten by an...

Octopus, octopus
Squiggling in the water.
Octopus, octopus
    Gulp, gulp, gulp

Oh, no!
It's been eaten by a...

Tuna fish, tuna fish
Flashing through the water.
Tuna fish, tuna fish
    Gulp, gulp, gulp

Oh, no!
It's been eaten by a...

Great white shark, great white shark
Lurking in the water.
Great white shark, great white shark
    Gulp, gulp, gulp

Oh, no!
It's been eaten by a...

Humongous whale, humongous whale
Spouting in the water.
Humongous whale, humongous whale
    Gulp, gulp, gulp!

    PARDON ME!

Use with **p. 13 Slippery Fish**

Ocean Animals

Which would you

broccoli

carrots

Which water bear from

Prediction

Hot

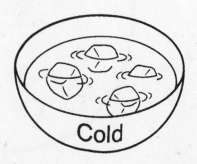

Cold

Use with **p. 15 What Would You Grow?**
Use with **p. 16 Wake Up Bear**

like to grow?

corn

will wake your hibernation?

Conclusion

Hot

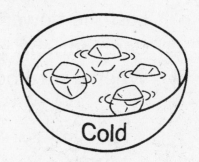

Cold

Use with **p. 15 What Would You Grow?**
Use with **p. 16 Wake Up Bear** 77

Germination Song

(Tune: "Found a Peanut")

Germination, germination,
When a seed begins to grow.
Germination's what we call it
When a seed begins to grow.

Germination, germination,
Next the roots begin to grow.
Germination's what we call it
When a seed begins to grow.

Germination, germination,
Next the stem begins to grow.
Germination's what we call it
When a seed begins to grow.

Germination, germination,
Next the leaves begin to grow.
Germination's what we call it
When a seed begins to grow.

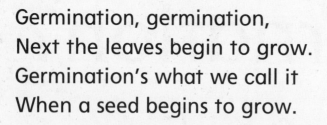

Germination, germination,
Last the flower begins to grow.
Germination's what we call it
When a seed begins to grow.

Use with **p. 14 Germination**

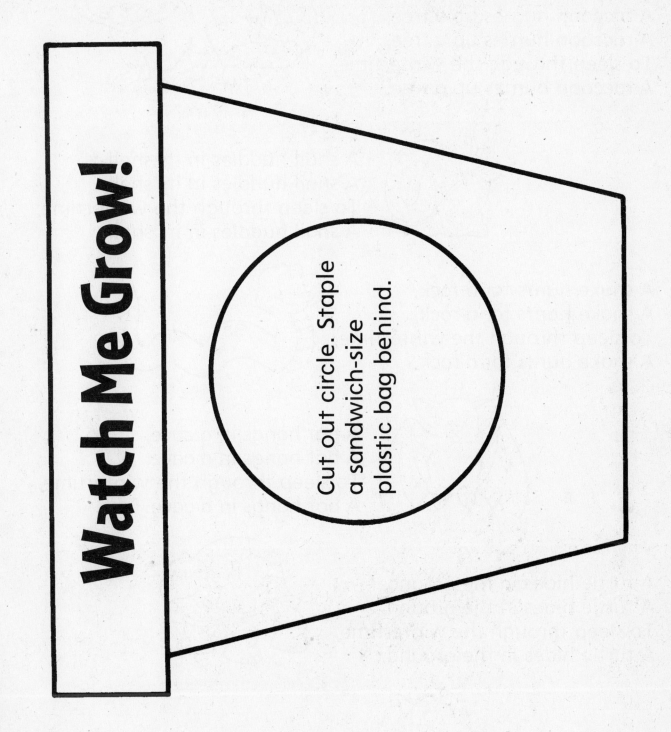

Watch Me Grow!

Cut out circle. Staple a sandwich-size plastic bag behind.

H is for Hibernation

(Tune: "Farmer in the Dell")

A raccoon hurries up a tree,
A raccoon hurries up a tree,
To sleep through the wintertime,
A raccoon hurries up a tree.

A snail huddles in its shell,
A snail huddles in its shell,
To sleep through the wintertime,
A snail huddles in its shell.

A snake hunts for a rock,
A snake hunts for a rock,
To sleep through the wintertime,
A snake hunts for a rock.

A bat hangs in a cave,
A bat hangs in a cave,
To sleep through the wintertime,
A bat hangs in a cave.

A turtle hides in the ground,
A turtle hides in the ground,
To sleep through the wintertime,
A turtle hides in the ground.

A frog hops in the mud,
A frog hops in the mud,
To sleep through the wintertime,
A frog hops in the mud.

Use with **p. 16 Hibernating Habitats**

Hibernating Animals

raccoon 1	snail 2
snake 3	bat 4
turtle 5	frog 6

Hibernating Habitats

1 a tree	2 a shell
3 a rock	4 a cave
5 the ground	6 the mud

Use with p. 16 Hibernating Habitats

Bear Facts

(Tune: "If You're Happy and You Know It")

We know that bears eat honey, nuts, and fish.
We know that bears eat honey, nuts, and fish.
They fill up their tummy
With things they think are yummy.
We know that bears eat honey, nuts, and fish.

We know that bears get furry in the fall.
We know that bears get furry in the fall.
To keep them warm and snug
So they sleep as tight as a bug.
We know that bears get furry in the fall.

We know that bears get sleepy when it's cold.
We know that bears get sleepy when it's cold.
They go into their den
And sleep till winter's end.
We know that bears get sleepy when it's cold.

We know that bears wake up in the spring.
We know that bears wake up in the spring.
They come out of their den
To start it all again.
We know that bears wake up in the spring.

Bear Pattern for "Bear Facts" Book

Copy on card-stock paper, color, cut out, and glue on craft stick.

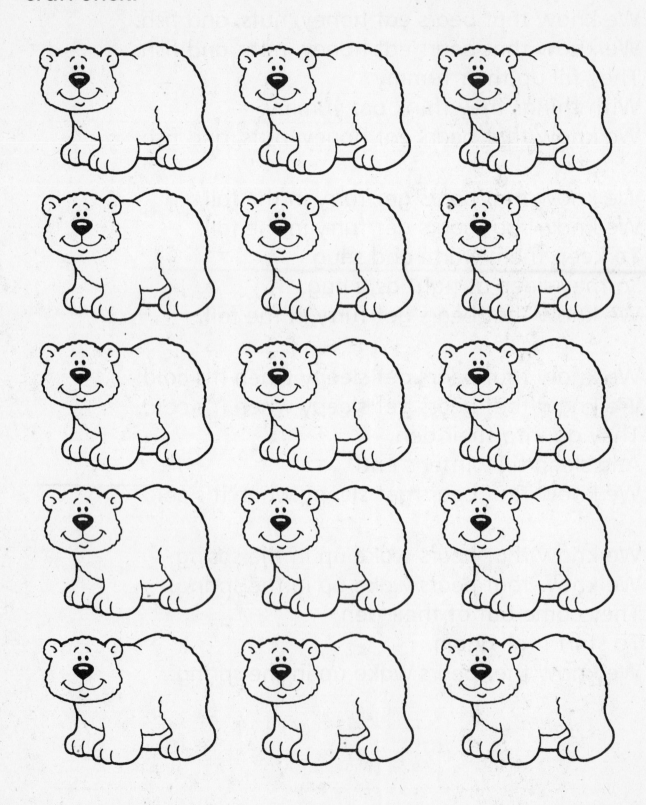

Use with **p. 17** "The Bear Facts"

Animals that hibernate and animals that do not hibernate	raccoon	turtle
bat	monkey	snake
cow	bird	mouse

Insects for Positional Chart

Use with **p. 18 Position Chart**

Logic Line-Up Directions

Problem #1

1. The insect that crawls is to the left of the one that hops.
2. The one with spots is not before the one that crawls.
3. The insect that crawls and the insect with stripes are on the ends.
4. The insect that hops is before the one with spots.

Problem #2

1. There are at least two insects before the insect with no wings.
2. The insect that produces honey is not in the middle.
3. The insect that eats aphids is beside the insect that makes honey.
4. The insect with no wings is before the insect with long back legs.

Problem #3

1. The insect that is red is not on an end.
2. The insect that changes into something else is not in the middle.
3. The insect that hops in the grass has two insects before it.
4. The insect that flies above flowers is to the left of the insect that sits on leaves.

Logic Line-Up Cards

Ladybug

Caterpillar

Bee

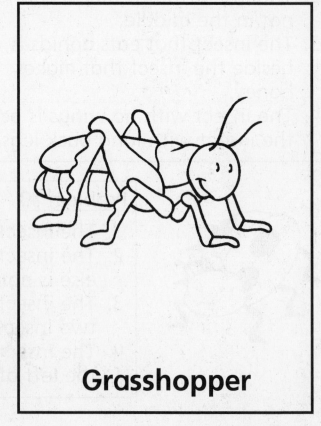

Grasshopper

Use with **p. 18 Baffling Bugs**

Irresistible Ladybugs

Color of Ladybug

What is your favorite insect?

Red circle — Ladybug
Yellow circle — Ant
Orange circle — Butterfly

Dots

Are you afraid of insects?

Yes — 2 dots
No — 4 dots

Antennas

Have you ever caught a ladybug?

Yes — curled antenna
No — straight antenna

Add Leaves

Would you rather be a flying insect or an insect that crawls?

Crawling — 1 leaf
Flying — 2 leaves

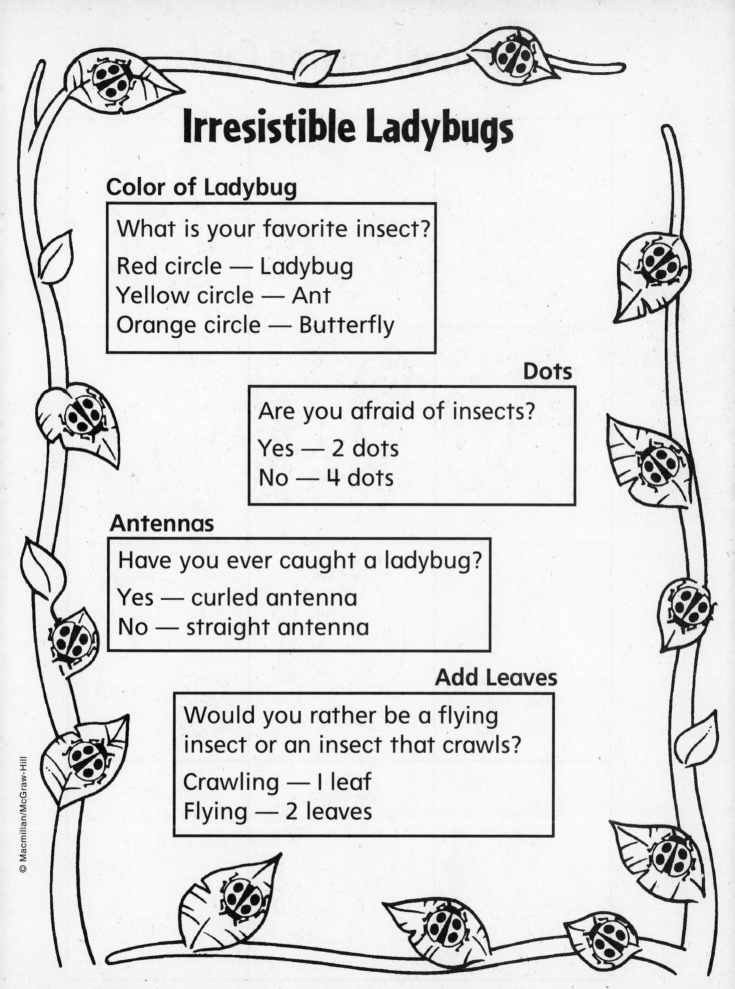

Animal Sorting Cards

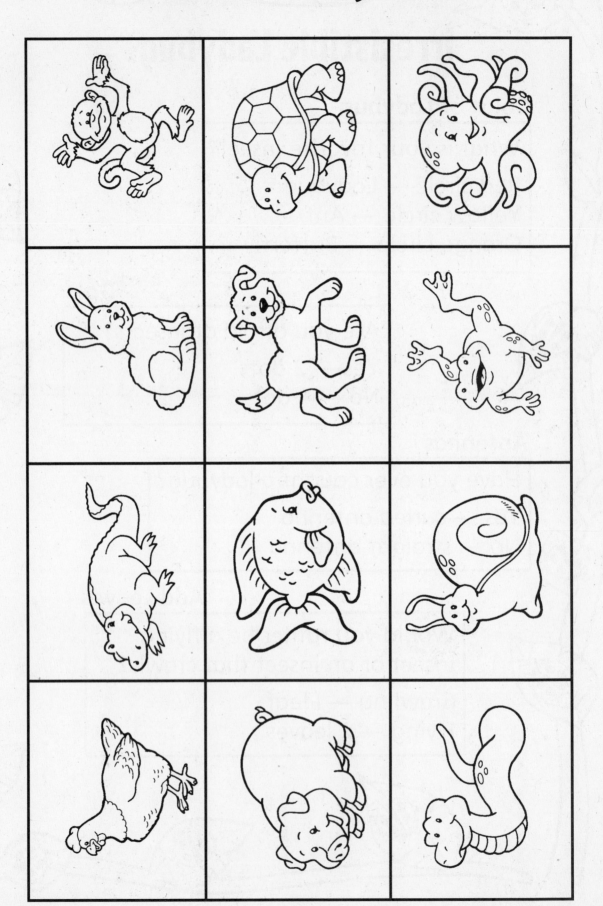

Use with **p. 20 Jump Up**

Jumping Frog

The frog jumps in the flowers.

The frog jumps by the tree.

The frog jumps into the water.

The frog jumps on the rock.

Name _____

Just Measure

_____ connecting cubes

_____ toothpicks

_____ beans

_____ counting bears

_____ linking loops

Directions: Use connecting cubes, toothpicks, beans, counting bears, and linking loops to measure the frog. Measure the frog from top to bottom.

Use with **p. 21 Just Measure**

Jump, Bunny, Jump!

by _____

Jump through the grass.

Jump on a rock.

Jump beside a flower.

Jump in a puddle.

Jump under a tree.

Jump back to his home.

Bunny for "Jump, Bunny, Jump!"

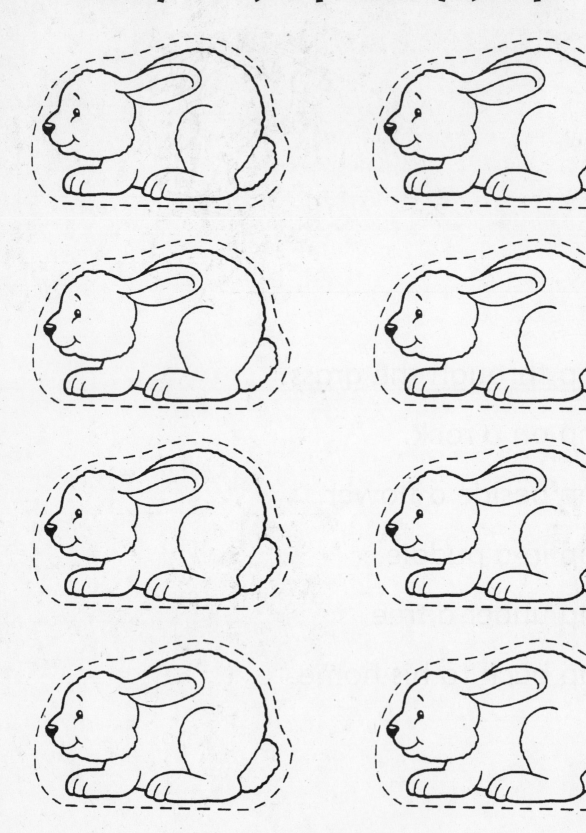

Use with **p. 21 Jumping Bunny**

Name _____

Kindergarten Kids' Fingerprints

Our fingerprints are unique, but they can usually be grouped into three categories. Put a finger on a stamp pad and push your finger in the box. Use a magnifying lens to determine if your fingerprint fits in one of these groups.

arch   loop   whorl

Put more fingerprints below and create pictures from them. Have fun!

Egg Activity Sheet

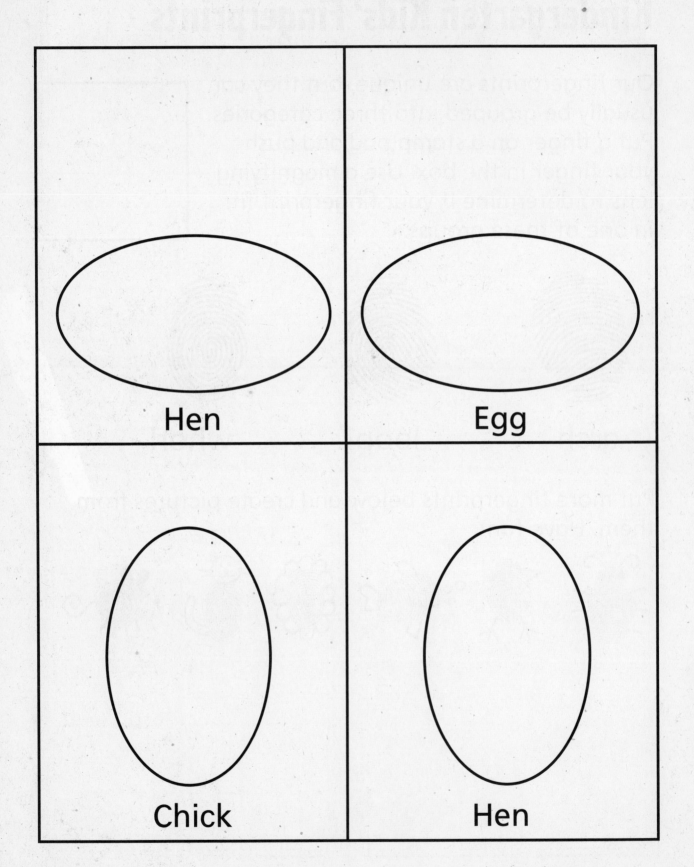

Hen

Egg

Chick

Hen

Use with **p. 25 Chicken Life Cycle**

Magnets

Check the objects to see if they are attracted to magnets.

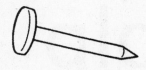 Yes

No

 Yes

No

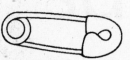 Yes

No

 Yes

No

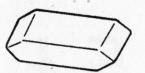 Yes

No

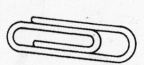 Yes

No

 Yes

No

More or Less

Weigh the objects in each row. Circle the items that weigh more.

1  clay ball	 crayons
2 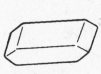 eraser	clay ball
3 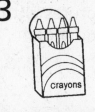 crayons	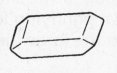 eraser
4 block	clay ball
5 scissors	block
6 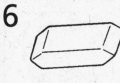 eraser	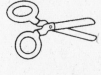 scissors

Magnifying Lens

Study a penny with a magnifying lens.

The year on the penny is _____ .

On the head of the penny I see...

On the tail of the penny I see...

Look at the small pictures with a magnifying lens and match to the large pictures.

Mirrors

Put a mirror next to words, read, and then write the words.

glad

cat

sad

Make letters with a mirror.

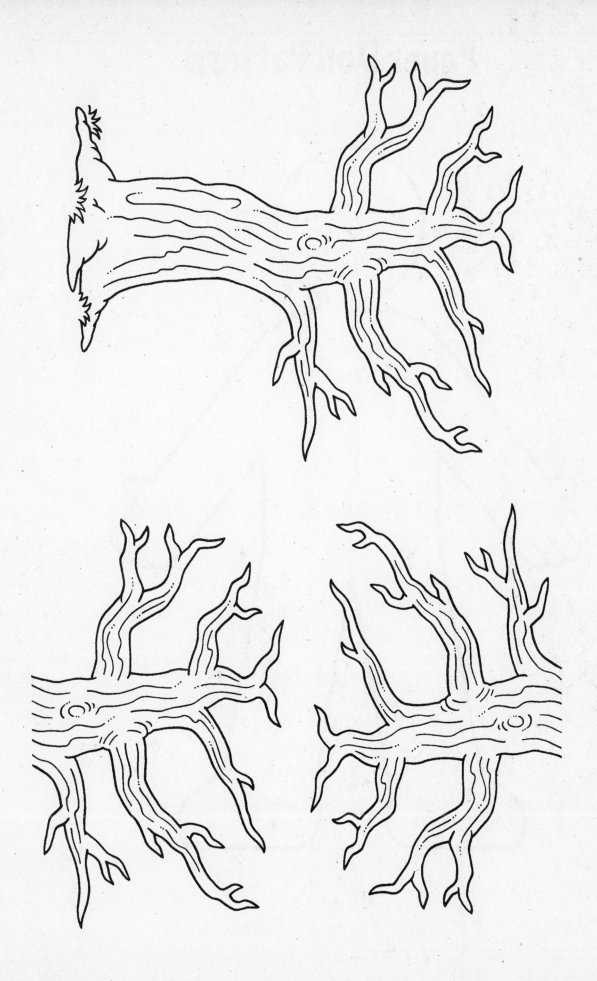

Paper Doll Pattern

Use with **p. 28 Nature and Me**

The Weather Song

(Tune: "Bingo")

What is the weather that you like?
Could it be sunny?
S-U-N-N-Y
S-U-N-N-Y
S-U-N-N-Y
Is sunny your favorite weather?

What is the weather that you like?
Could it be cloudy?
C-L-OU-D-Y
C-L-OU-D-Y
C-L-OU-D-Y
Is cloudy your favorite weather?

What is the weather that you like?
Could it be windy?
W-I-N-D-Y
W-I-N-D-Y
W-I-N-D-Y
Is windy your favorite weather?

What is the weather that you like?
Could it be rainy?
R-A-I-N-Y
R-A-I-N-Y
R-A-I-N-Y
Is rainy your favorite weather?

What is the weather that you like?
Could it be snowy?
S-N-O-W-Y
S-N-O-W-Y
S-N-O-W-Y
Is snowy your favorite weather?

Opposites: Sink or Float

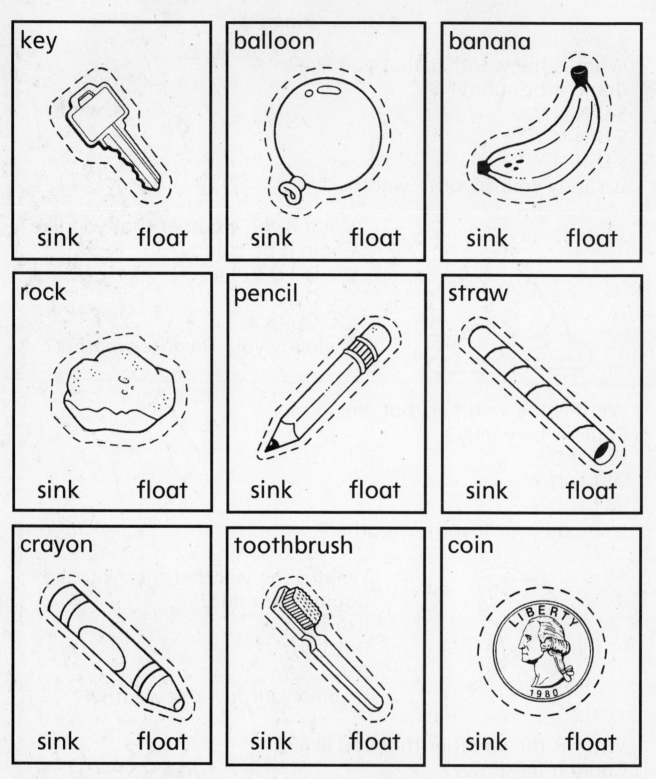

key	balloon	banana
sink float	sink float	sink float

rock	pencil	straw
sink float	sink float	sink float

crayon	toothbrush	coin
sink float	sink float	sink float

Test each item in a tub or bin of water to see if it sinks or floats. Circle the answer. Then cut out and glue the items on the Sink/Float sorting mat.

Use with **p. 30 Sink or Float**.

float

sink

Use with **p. 30 Sink or Float** 103

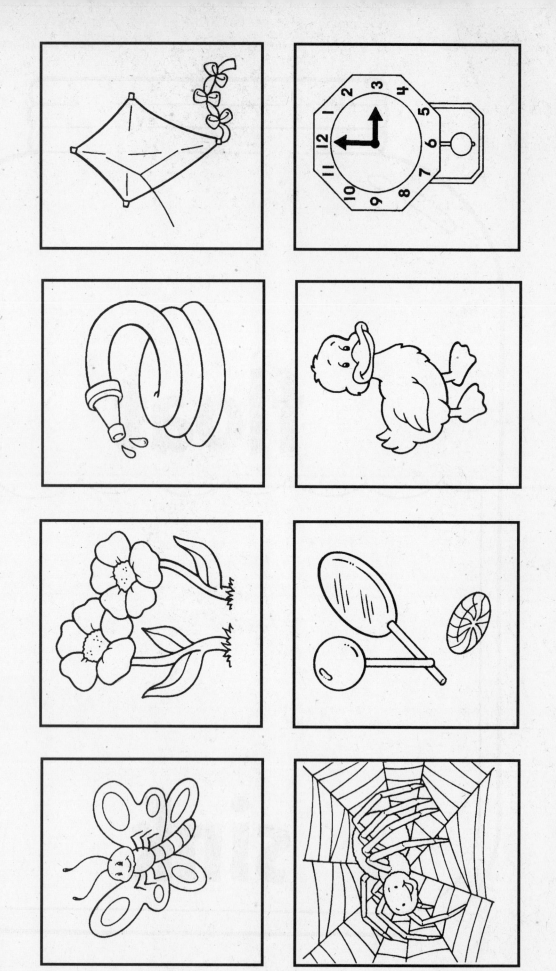

Opposites: Living and Nonliving Things

Cut out and sort the pictures to show living and nonliving things.

Opposites: Living and Nonliving Things

Cut out and sort the pictures to show living and nonliving things.

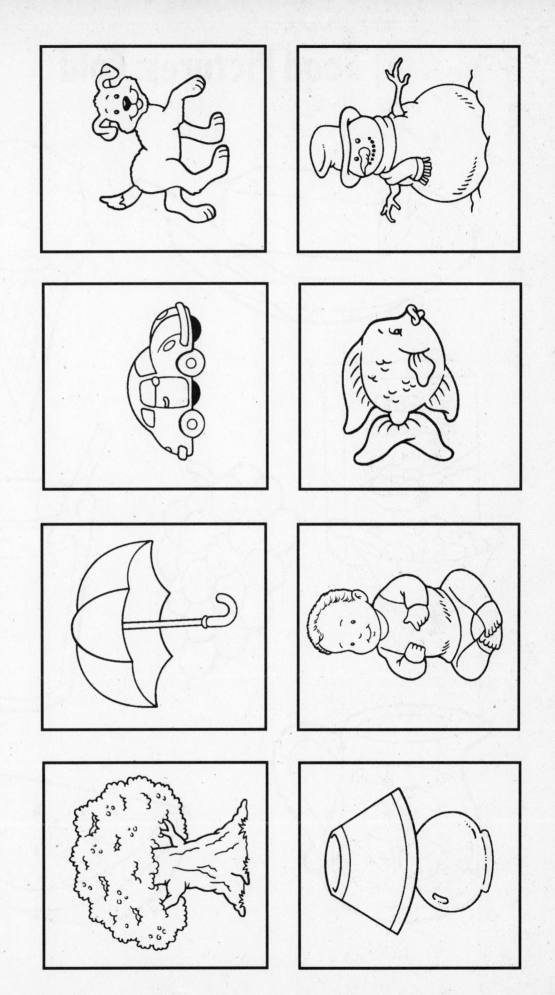

Food Pictures: Cold

SUPER FLAKES

Food Pictures: Hot

Opposites: Healthy and Unhealthy Habits

Take turns choosing role playing cards and acting out the situations. Discuss if the behaviors are healthy or unhealthy habits.

Washing hands with soap.	Doing work and then not cleaning hands.	Going to bed at a regular time.	Not going to bed at a regular time.
Playing and having fun with others.	Getting mad when playing with others.	Getting lots of exercise.	Watching lots of television.

Use with p. 31 Healthy and Unhealthy

Opposites: Healthy and Unhealthy Habits

Solving problems with words and good actions.	Fighting to solve problems.	Not caring for an injury.	Covering an injury with a bandage.
Eating junk food regularly.	Eating healthy food regularly.	Causing injuries by poor habits.	Staying safe with good habits.

How far did your

Prediction

0–10　　11–20　　21–30　　31+

How many _____

Prediction

0–5　　6–10　　11–15　　16+

Use with **p. 32 Paper Planes**
Use with **p. 32 Powerful Paper**

plane soar?

Conclusion

0–10   11–20   21–30   31+

will my boat hold?

Conclusion

0–5   6–10   11–15   16+

Use with p. 32 Paper Planes
Use with p. 32 Powerful Paper 111

Use with p. 32 Paper Planes

 # Paper Predictions

What kind of paper absorbs water?

Paper	Prediction	Conclusion

People Poppers

Use with p. 33 People Popping

Oh, Can You Hear?
(Tune: "Muffin Man")

Oh, can you hear a flower bloom,
A flower bloom, a flower bloom?
Oh, can you hear a flower bloom?
No, oh no, I can't.

Oh, can you hear a dog barking,
A dog barking, a dog barking?
Oh, can you hear a dog barking?
Yes, oh yes, I can.

Oh, can you hear the grass growing,
The grass growing, the grass growing?
Oh, can you hear the grass growing?
No, oh no, I can't.

Oh, can you hear a clock ticking,
A clock ticking, a clock ticking?
Oh, can you hear a clock ticking?
Yes, oh yes, I can.

Oh, can you hear a car go by,
A car go by, a car go by?
Oh, can you hear a car go by?
Yes, oh yes, I can.

Oh, can you hear a butterfly,
A butterfly, a butterfly?
Oh, can you hear a butterfly?
No, oh no, I can't.

In box 2 there is...

In box 5 there is...

In box 3 there is...

In box 1 there is...

In box 4 there is...

Use with **p. 35 Quiet, Listen, and Guess**

I'm a petrologist

Rock Sorting Labels

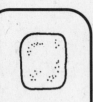 **square**

 **rough**

 striped

 **oval**

 **speckled**

 **shiny**

 **not shiny**

 **bumpy**

 small

 round

Use with **p. 36 Rock Sorting**

Moon Rocks

Ingredients (Check children's allergies):

$\frac{1}{2}$ cup peanut butter $\frac{1}{2}$ cup powdered sugar

$\frac{1}{2}$ cup dry milk $\frac{1}{2}$ cup cereal

4 oz. semisweet chocolate, melted   2 T. butter, melted

Procedure:

1. Wash hands. Mix peanut butter, sugar, dry milk, and cereal together.
2. Add melted butter and chocolate, then mix.
3. Roll dough into a ball.
4. Roll in powdered sugar.
5. Put them on a foil-covered plate.
6. Refrigerate until firm. Store in covered container.

Edible Rocks

Ingredients (Check children's allergies):

$\frac{1}{4}$ cup peanut butter

2 T. powdered sugar

2 T. sweetened condensed milk

Procedure:

1. Wash hands. Put all ingredients in a bowl.
2. Mix well.
3. Take a pinch of mixture and roll into a ball.

How many rocks did you make?

Does it shine through?

Object	Prediction	Conclusion
aluminum foil	Yes No	
wax paper	Yes No	
paper towel	Yes No	
cardboard	Yes No	
clear plastic bag	Yes No	

Use with **p. 39 Shine Through**

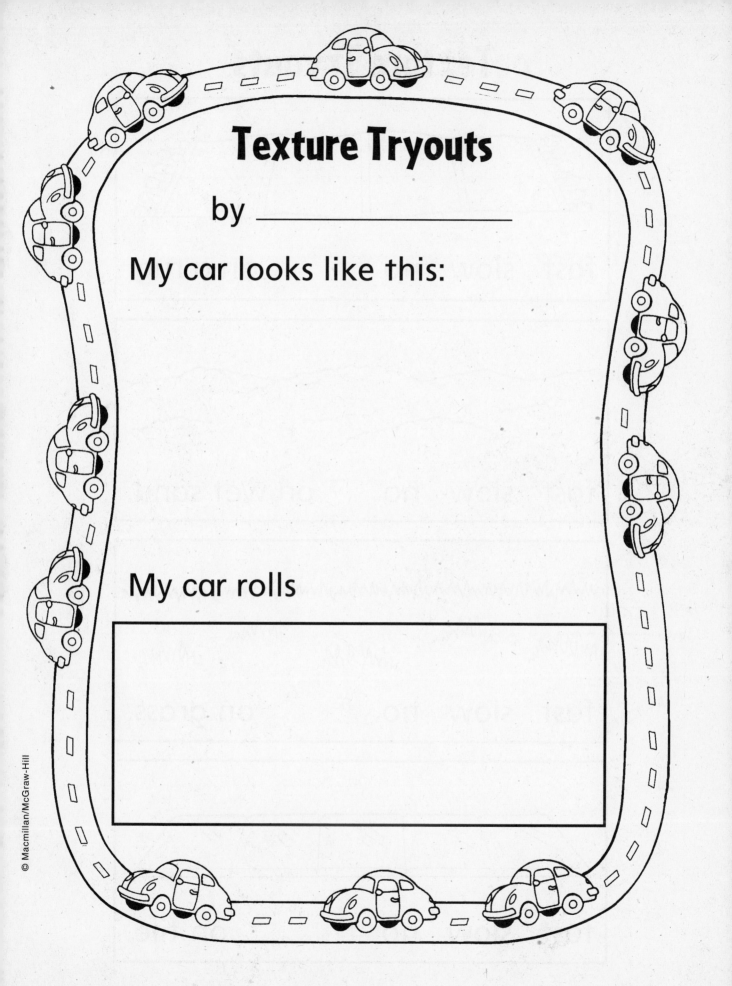

Texture Tryouts

by _____

My car looks like this:

My car rolls

Texture Tryouts

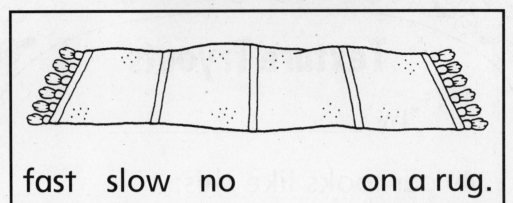

fast   slow   no   on a rug.

fast   slow   no   on wet sand.

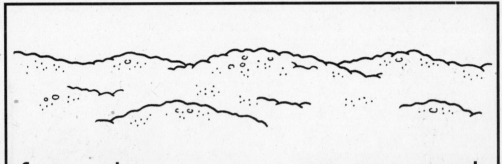

fast   slow   no   on grass.

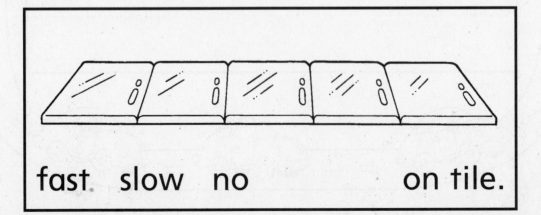

fast   slow   no   on tile.

Use with **p. 40 Texture Tryouts**

Texture Tryouts

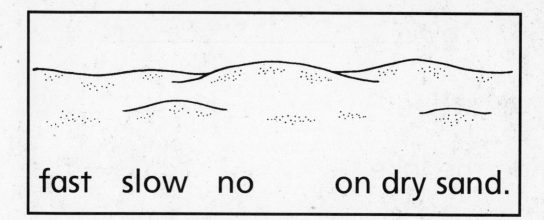

fast   slow   no   on dry sand.

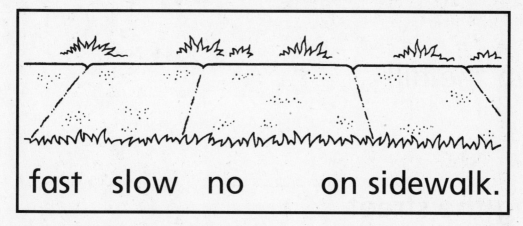

fast   slow   no   on sidewalk.

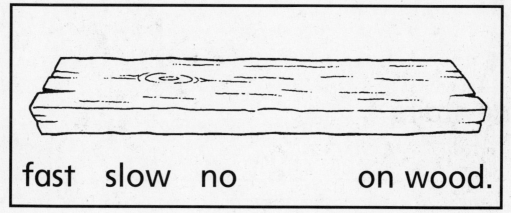

fast   slow   no   on wood.

Wheel Away

by _____

beside the lake

down the hill

along the street

up the stairs

under the _____.

Can You Guess How People Move?

With four wheels and a motor,
You can go really far!
Buckle up and then ride in
This thing called a _____.

When you put up the sail,
Over water you will float.
Climb aboard and sail off in
This thing called a _____.

This takes children to school,
Holds a lot without a fuss.
Load up and ride off in
This thing called a _____.

Your two legs will move you,
But you won't need to hike.
Hop up and ride off on
This thing called a _____.

It can fly through the sky,
Through the wind and the rain.
Climb aboard and take off in
This thing called a _____.

It walks and it runs,
And it gallops, of course!
Take a ride on the back of
This thing called a _____.

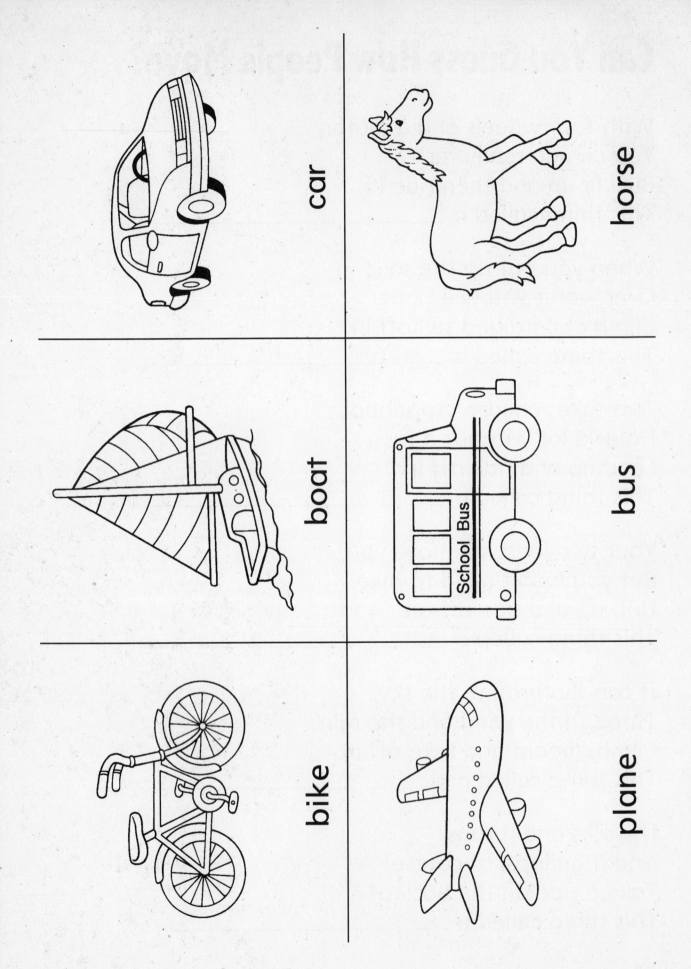

car

horse

boat

bus

School Bus

bike

plane

Use with **p. 41 Transportation Trivia**

Let's Go Traveling

(Tune: "Row, Row, Row Your Boat")

Sail, sail, sail the boat,
Gently 'round the lake,
Moving, moving on the water,
Boats are what you take.

Ride, ride, ride the bus,
Have a merry cruise.
Moving, moving on the road,
Buses are what you use.

Float, float, float the balloon,
Way up in the sky.
Moving, moving in the clouds,
Balloons go way up high.

Chug, chug, chug, the train,
Rolling across the road.
Moving, moving down the rails,
Trains carry such a load.

Fly, fly, fly the plane,
High up in the air.
Moving, moving through the sky,
Planes can get you there.

Stamp, stamp, stamp your feet,
Stamp them on the ground.
Moving, moving on your feet,
Walk to get around!

© Macmillan/McGraw-Hill

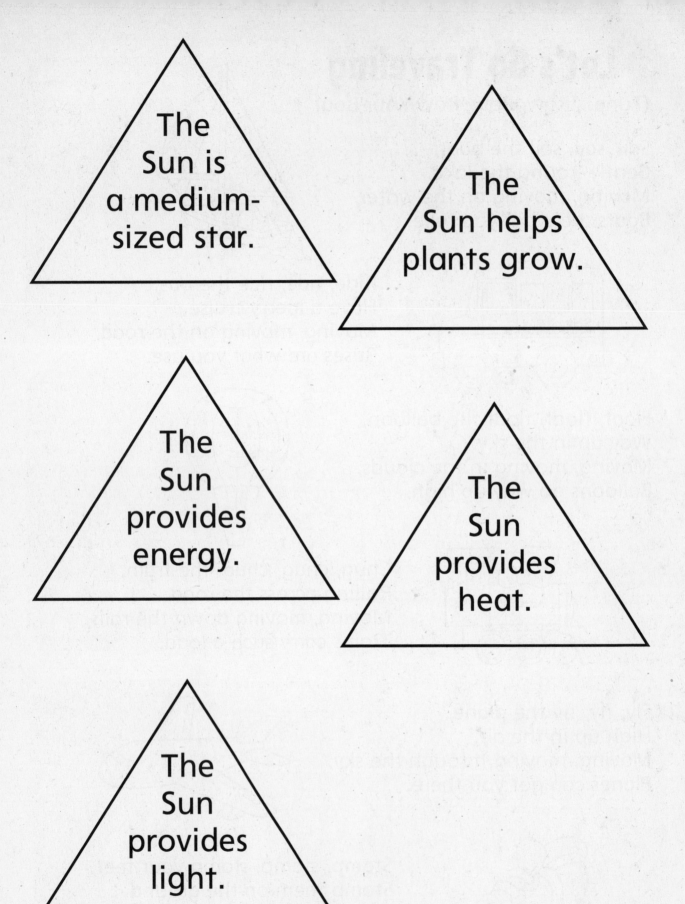

The Sun is a medium-sized star.

The Sun helps plants grow.

The Sun provides energy.

The Sun provides heat.

The Sun provides light.

Use with **p. 42 The Sun**

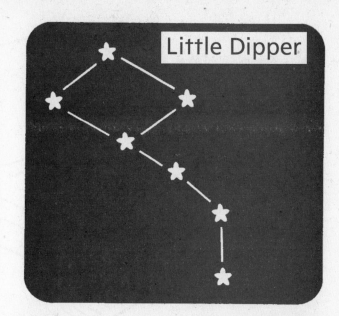

Big Dipper

Little Dipper

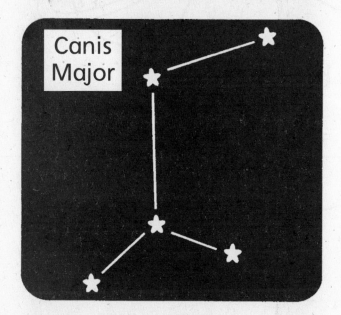

Canis Major

Canis Minor

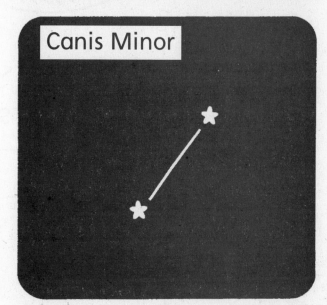

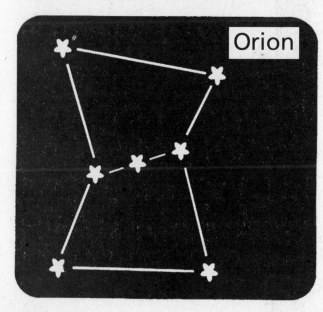

Orion

Cassiopeia

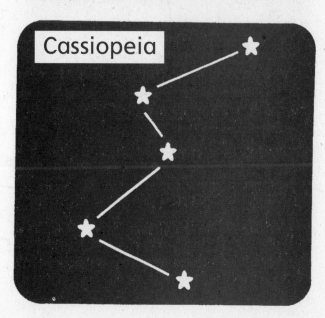

© Macmillan/McGraw-Hill

Planets big,
planets _____.

Some planet names
I can't recall.

Use with p. 43 Planets

Planets hot,
planets _____.

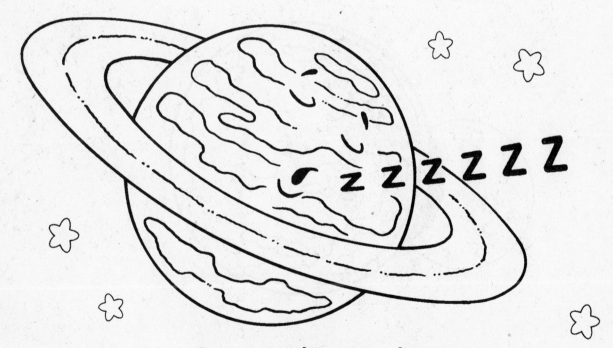

Some planets have
no life, I'm told.

Planets here,
planets _____.

I wonder if more planets
are hiding somewhere.

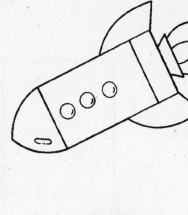

Unique Shapes in the Universe

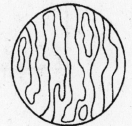

by _____

I see a space shuttle.
_____ shuttle,
_____ shuttle,
I see a space shuttle.

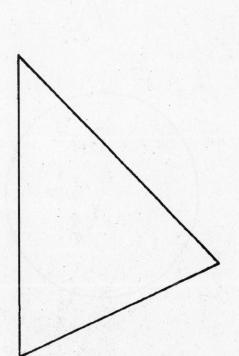

I see a rocket ship.

I see a _____ ship,

_____ ship,

I see a rocket ship.

I see a planet.

I see a _____ planet,

_____ planet,

I see a planet.

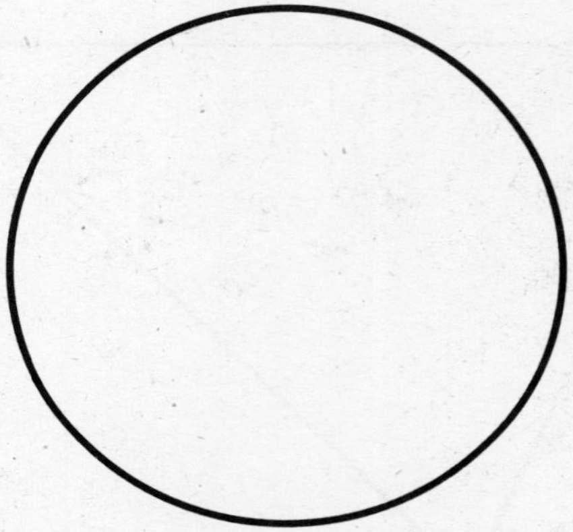

Use with **p. 43 Unique Shapes**

Space Pictures

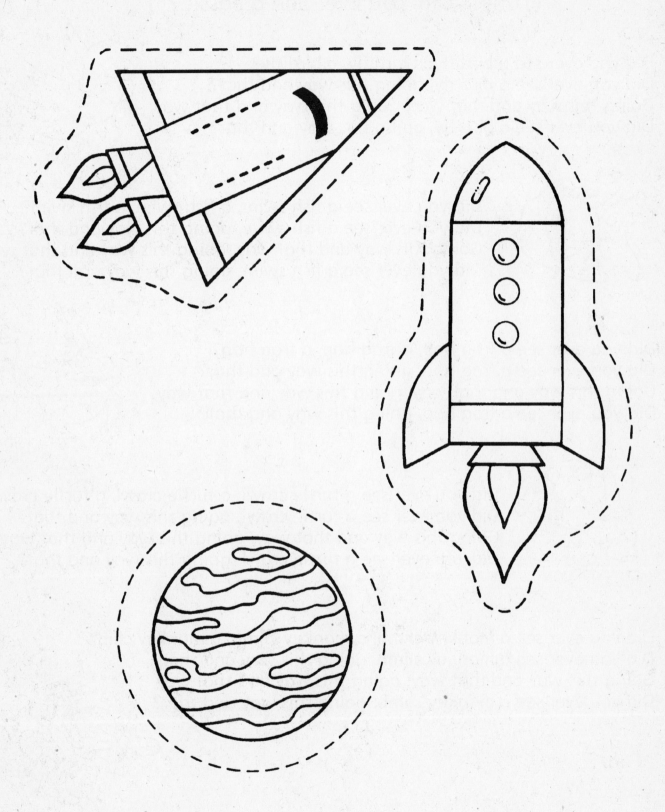

Did You Ever See?

(Tune: "Did You Ever See a Lassie?")

Did you ever see a bird fly, a bird fly, a bird fly?
Did you ever see a bird fly, going this way and that?
Going this way and that way, going this way and that way,
Did you ever see a bird fly, going this way and that?

Did you ever see a fish swim, a fish swim, a fish swim?
Did you ever see a fish swim, going this way and that?
Going this way and that way, going this way and that way,
Did you ever see a fish swim, going this way and that?

Did you ever see a frog hop, a frog hop, a frog hop?
Did you ever see a frog hop, going this way and that?
Going this way and that way, going this way and that way,
Did you ever see a frog hop, going this way and that?

Did you ever see a turtle crawl, a turtle crawl, a turtle crawl?
Did you ever see a turtle crawl, going this way and that?
Going this way and that way, going this way and that way,
Did you ever see a turtle crawl, going this way and that?

Did you ever see a monkey climb, a monkey climb, a monkey climb?
Did you ever see a monkey climb, going this way and that?
Going this way and that way, going this way and that way,
Did you ever see a monkey climb, going this way and that?

Use with **p. 44 Eye Color Chart**

Name _____

Eye Color Chart

brown	**green**	**blue**

Optical Illusions

Use with **p. 44 Optical Illusions**

Optical Illusions

The rain fills the river.
The river flows into the ocean.

The Sun heats the ocean.

Use with **p. 46 The World of Water**

Water goes into the clouds.

**The clouds get big and dark,
and down comes the rain again.**

Water Wiggle

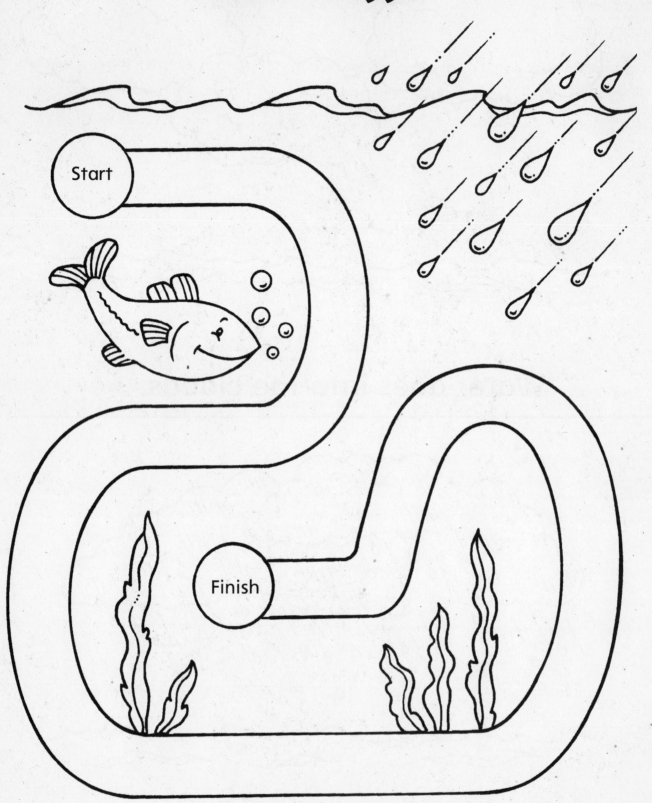

Start

Finish

Use with **p. 47 Water Wiggle**

"X-tra" Special Me!

Directions: Using crayons, decorate the paper doll to look like yourself. Hold it up to the light and see the X-ray of "X-tra" special you!

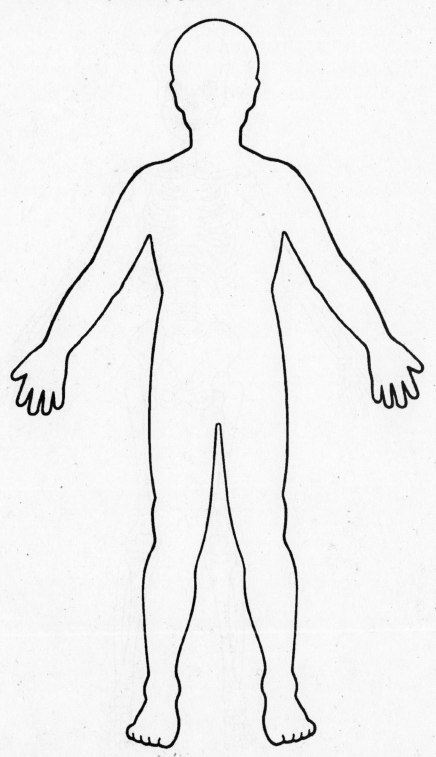

"X-tra" Special Me!

Directions: ...

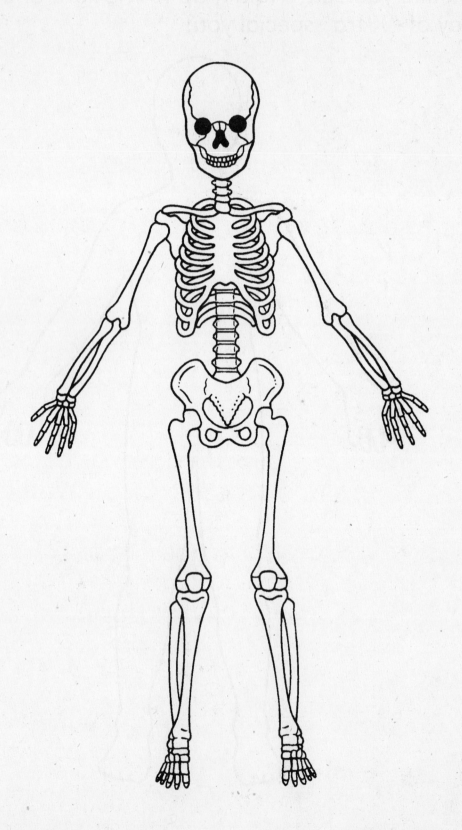

Use with **p. 48 "X-tra" Special Me!**

Them Hand Bones

(Tune: "Dry Bones")

The **thumb** bones are connected to the **pointer** bones.
The **pointer** bones are connected to the **tall** bones.
The **tall** bones are connected to the **ring** bones.
Last of all the **pinky** bones.

Them bones, them bones,
them hand bones.
Them bones, them bones,
them finger bones.
Them bones, them bones,
them hand bones.
We use them all day long.

We **shake** and **clap** and **write** with them.
We **squeeze** and **hold** and **cut** with them.
We **color**, we **throw** and **catch** with them.
They **bend**, they **stretch**, they **rest**.

Them bones, them bones,
them hand bones.
Them bones, them bones,
them finger bones.
Them bones, them bones,
them hand bones.
We need to keep them strong!

© Macmillan/McGraw-Hill

"X-amine X-tinct" Creatures

How many bones?

_____

Use with p. 49 "X-tinct" Creatures

This is a _____.

This is a _____.

This is a _____.

This is a _____.

Use with **p. 49 Animal X-Rays**

Young Animals and Their Names

A baby sheep is a lamb.

A baby cat is a kitten.

A baby bear is a cub.

A baby cow is a calf.

A baby dog is a puppy.

A baby horse is a foal.

A baby duck is a duckling.

A baby pig is a piglet.

Young Animals and Their Names

lamb

 kitten

cub

 calf

puppy

 foal

duckling

 piglet

Use with **p. 50 Young Animal Names**

I am old enough to

_____,

but I am too young to

_____.

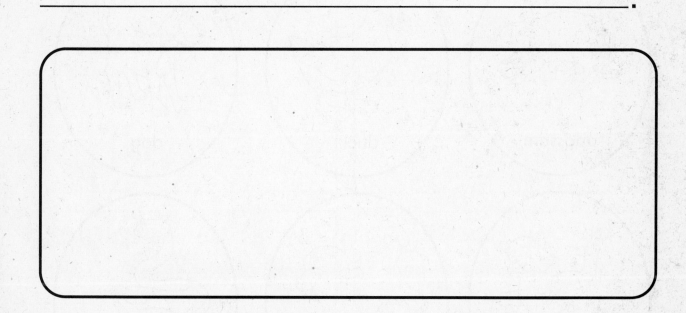

by _____

bat

alligator

owl

cow

coyote

lion

opossum

duck

dog

cat

lamb

rattlesnake

Use with p. 52 Diurnal or Nocturnal?

Nighttime Noises

(Tune: "This Little Light of Mine")

I hear a noise at night.
It's a flapping bat. Whoosh!
I hear a noise at night.
It's a flapping bat. Whoosh!
What's it doing at night?
It's chasing a bug.
Hear it flap, hear it flap, hear it flap!

I hear a noise at night.
It's an old hoot owl. Hoot!
I hear a noise at night.
It's an old hoot owl. Hoot!
What's it doing at night?
It's chasing a mouse.
Hear it hoot, hear it hoot, hear it hoot!

I hear a noise at night.
It's an old raccoon. Splash!
I hear a noise at night.
It's an old raccoon. Splash!
What's it doing at night?
It's chasing a fish.
Hear it splash, hear it splash, hear it splash!

© Macmillan/McGraw-Hill

The Bear and the Moon

___

The bear is <u>on</u> the Moon.

___

The bear is <u>under</u> the Moon.

___

The bear is <u>above</u> the Moon.

___

The bear is <u>beside</u> the Moon.

___

The bear is <u>below</u> the Moon.

___

The bear is <u>behind</u> the Moon.

___

The bear is <u>in front of</u> the Moon.

___

Bear Patterns

Use with **p. 53 Where Is the Bear?**